T0010579

BUILD YOUR OWN
FARM TOOLS

JOSH VOLK

Storey Publishing

Dedicated to the kids of all ages who take their toys apart and try to put them back together again.

The mission of Storey Publishing is to serve our customers by publishing practical information that encourages personal independence in harmony with the environment.

Edited by Deborah Burns and Mia Lumsden
Art direction and book design by Erin Dawson
Indexed by Samantha Miller
Cover photography by © Shawn Linehan, except author by © Josh Volk
Cover and interior illustrations by © Michael Gellatly, except Ilona Sherratt © Storey Publishing, LLC, 15, 28 top, 33, 53, 57, 67, 80 bottom, 89, 142, 158, 186, 187, 194 right, and © Josh Volk, 173

Be sure to read all the instructions thoroughly before undertaking any of the projects in this book and follow all of the safety guidelines summarized on pages 8-9.

Storey books are available at special discounts when purchased in bulk for premiums and sales promotions as well as for fund-raising or educational use. Special editions or book excerpts can also be created to specification. For details, please call 800-827-8673, or send an email to sales@storey.com.

Storey Publishing
210 MASS MoCA Way
North Adams, MA 01247
storey.com

Printed in the United States by Bradford & Bigelow
10 9 8 7 6 5 4 3 2 1

Library of Congress Cataloging-in-Publication Data on file

CONTENTS

INTRODUCTION

WHY BUILD YOUR OWN TOOLS?

In this day and age you can buy a staggering array of tools for seemingly every purpose, so why would you want to create your own—especially when you can simply order what you need and have it delivered to your doorstep? On the farm we collect tools of all sizes and shapes to make our work easier, and many of us have toolboxes full of simple tools (like screwdrivers and wrenches) we use to fix the more complex tools (like carts and irrigation systems). A lot of us also use the tools we already have to build new tools and customize old ones because, despite appearances, you can't really buy a tool that fits every individual need. It's also rewarding to make useful things with our own hands; I always appreciate using the tools I've built for myself more than the ones I've purchased.

I love tools. I love learning how to use tools, I love modifying tools to make them work better for my own needs, and I love creating new tools. The tools I prefer are in some ways a reflection of my personality and the way I like to work. They are simple, functional, and efficient. For me, form follows function, and while my tool designs may have some small aesthetic flourishes, those flourishes never take away from the tool's essential functions or add significantly to its cost.

How This Collection Came to Be

This book features a collection of tools that I've built for use on my own farm and on friends' farms. All of the 19 projects that follow are tools that have seen extensive use and whose designs have worked well for me and for my farmer friends. These tools are all designed around commonly available materials and can be built with commonly available tools on the farm.

The book starts with a chapter on the basic tools and materials you'll need to create the new tools that make up the rest of the book. Each building project includes a list of the supplies you'll need and step-by-step instructions. Each project also features notes about how I use the tool as a part of a system on the farm. Without this context, it's not always obvious why a tool works.

I also provide notes on design considerations for modifying or customizing the tools, or perhaps making future improvements. These notes on use and design are meant to encourage you to modify these designs to better meet your farm's needs. In addition, the notes help explain some of the "whys" behind the designs. In many cases these notes point out design features that I came to after trying other methods that were less successful.

Not all of these tools are completely my own designs. In fact, none of the tools truly are "mine," in the sense that they were all inspired by tools that I've seen on farms or in other places. In some cases, I remember where the idea for a tool came from; in other cases, I've been using a tool for so long or it's morphed so much that I can't fully trace its origins. These designs are mine, however, in the sense that I've figured out my own ways to construct them and use them.

I hope that you will build these tools and find them useful—and that you'll modify them, if needed, to better suit your own needs. I also hope that you'll see some of these tools, realize they're not useful in your particular situation, but still maybe note the ways in which they are put together and be able to apply those design features to your own future tool designs. The information in Chapter 1, the appendix, and the tips and design notes scattered throughout the book are all there to help you with tweaks, modifications, and ground-up design. The way we all learn to build better tools is by first observing

existing designs, then interacting with them, and then trying new methods.

When we use a tool, we learn about what we like and dislike about it; that should inspire experimentation with modifying existing tools and designing new ones. From a learning perspective, the aspects of a tool that don't work for you are just as important as those that do. As you read through the following pages, I encourage you to be willing to make your own mistakes and learn from them as much as you do the directions you'll find here.

CHAPTER 1
SETTING UP A BASIC SHOP

IT'S NOT NECESSARY TO HAVE A DEDICATED SHOP SPACE to start building or modifying tools, but at some point having a shop makes it easier to get started on a project because you don't have to pull tools from the back of a closet and set up a work surface every time you want to build something or tinker. I've found that, over time, items such as basic fasteners, extra bits of wood and metal, and random raw materials for future projects start to collect and need a place to live, and that space is usually a dedicated shop.

I'm a fan of starting small and keeping things relatively simple. I avoid buying specialized tools that I don't absolutely need and will use only for a single foreseeable project, even if they're a good deal. On the other hand, I have a few tools that I use consistently and recommend for every small-farm shop. If you don't have them already, you don't need to go out and buy them all at once, but when you do need them, they're worth purchasing.

General Tool Safety Notes

BEFORE YOU START WORKING with any tool, think about the physical space you'll be working in. It pays to take the time to clear out a sufficient workspace, making sure you can stand properly, hold the work properly, and view it under proper lighting. Ultimately this will save you time and frustration, while also preventing injuries and wear and tear on your body.

When working with shop tools, consider your attire. Wear gloves to protect your hands, closed-toed shoes to protect your feet, and long pants and long sleeves to protect your skin. You may also need eye protection, hearing protection, and other safety equipment such as a mask, a respirator, and even a hat or helmet depending on the tool you're using.

While it may seem like all this safety equipment could slow you down or create a less-safe situation by restricting your movement or vision, each of these items has saved me from serious injury on at least one occasion. Additionally, instead of slowing me down, they actually help me focus on the task and proceed with confidence.

Set up your workspace so you can move around comfortably and avoid awkward positions. If you're using a sharp tool, give yourself enough room so that if the tool slips, it won't have the path or momentum to cut you. In addition, secure heavy objects to prevent them from falling on your hands, feet, or body.

When welding or working with a grinder or anything else that creates sparks, think about fire safety and your own safety. Those sparks are small pieces of very hot metal that you want to keep away from your body and far away from anything they might ignite. Finally, when you're working around power cords, be aware of where they are so you don't cut them, and position them out of the line of traffic to prevent anyone from tripping over them. In general, think about safety as much as you think about the tool you're trying to build.

Proper lighting is an often-underappreciated safety feature around workshops and indoor workspaces on farms. In the short term, being able to see clearly what you're doing helps speed up tasks and improves accuracy and safety. In the longer term, working in well-lit spaces contributes to better productivity by reducing eyestrain and general fatigue.

The level of the light, the direction it comes from, and the color of both the light and the surroundings it bounces off all influence how well you will be able to see your work. Indirect natural ambient light combined with a bright light directly focused on your work is best, as it cuts down on glare. Walls and ceilings painted light colors will help brighten the workspace by allowing the light to bounce. Try to avoid the extreme contrast of working with a bright light focused on your work in an otherwise dark workspace.

TIP GLOVE SAFETY

Do not wear gloves if there's any chance that a glove might get snagged by a rotating blade and pull your hand into the blade. This is especially a worry if your gloves are oversized or if you need to work with your hand anywhere close to a rotating blade or shaft.

LOOKING CLOSER

POWER TOOL SAFETY

When you're working with power tools, I recommend taking the following precautions:

- **ALWAYS WEAR SAFETY GLASSES** to prevent splinters or chips from getting in your eyes.

- **KEEP YOUR FACE AWAY** from the path of flying sawdust or metal chips.

- **WEAR EAR PROTECTION** when using loud tools—basically any tool other than a cordless drill. I prefer the soft foam ear plugs that are held in place with a plastic hoop, but there are many good styles out there.

- **WEAR GLOVES** except in situations where the gloves may be counterproductive (see Glove Safety on the facing page). I usually wear form-fitting nitrile-dipped Atlas 370 gloves, which are thin but durable and offer a lot of dexterity as well as pretty good resistance to cuts and abrasion. I wear heavier leather gloves when working with objects that might be hot and when dexterity is less important.

- **SECURE LOOSE CLOTHING**, necklaces, bracelets, and long hair. The high-speed rotation of power-tool parts causes air movement that can suck loose items into their vicinity. They spin so fast that once something gets caught it takes only a fraction of a second for it to get tightly wrapped up, violently pulling you into the tool. Securing loose clothing and objects could literally save your life, or at least prevent unnecessary injuries.

- **WEAR AN APPROPRIATE MASK** and work in a well-ventilated area if you're dealing with any type of dust or fumes. Whether the dust is toxic is irrelevant; getting any quantity of small particles in your lungs is harmful. Disposable N95 masks are good to have on the farm for this purpose, and if you find yourself working around dust more often or for longer periods of time, you might consider a respirator with a better fit and replaceable cartridges (and only a respirator with the appropriate cartridges can protect you from fumes).

Basic Mechanic's Tools

ON THE FARM, adjustable wrenches, basic screwdrivers, and pliers are the primary tools I use for quick repairs. My first-to-buy list for an empty farm shop would be:

- 6-inch and 12-inch **adjustable wrenches.** The sizes indicate the length of the handle, which corresponds to the relative size of the hex head it will turn.

- 12-inch **adjustable pliers**. There are lots of styles, and I usually use the tongue-and-groove ones, but others work equally well.

- **Phillips and flat-head screwdrivers**—a basic set of two or three each in a variety of sizes and lengths. You can also get screwdrivers with replaceable tips. If you need tips for screws other than Phillips heads and flat heads, a multi-bit screwdriver will be helpful.

These are the mechanic's tools I find myself turning to constantly, and on the farm I also use them frequently for plumbing projects. I usually buy them from my local hardware store, not because their tools are the cheapest or the highest quality, but because the store is convenient and I like to support a local business that can help me answer questions when I have them. If the wrenches and pliers are at least decent quality and you take good care of them, they'll last forever.

Another tool that I use quite a bit is a **wire cutter**. There are many types, but what's most important is matching the size of the cutter to the size of the wire being cut. Trying to cut large wire with a small cutter is both difficult for you and damaging to the tool. Likewise, large cutters don't work on very small wire. On the farm, the wire shears incorporated into basic slip-joint pliers (not 12-inch adjustable pliers) work well for cutting baling wire and fencing pliers work for larger fence wire. If you're cutting a lot of heavy-gauge wire, you might want to use small bolt cutters, which require a little less hand strength.

Basic Woodworking Tools

IT IS GENERALLY EASIER to start working with wood than with metal, since wood is softer, less expensive, and more commonly available. Many basic woodworking tools can also be used with plastic (which often has a similar hardness to wood), and some do work with metal.

Drilling and Driving

These days **cordless drills** and basic sets of **drill bits** are widely available and relatively inexpensive. There's no one right size or make of drill to get, but you should think about how you are most likely to use the drill before purchasing a new one. In general, the less expensive the drill, the less durable it will be. More expensive drills will most likely last longer, have better features, and be more repairable if they break. Another consideration is weight: Lighter, lower-voltage cordless drills are less powerful, but they're also easier to carry and to use in awkward locations. Heavier, higher-voltage drills will run longer on a charge and offer more power, which is helpful if you need to drill big holes—but they're heavy!

TIPS DRIVING SCREWS

SCREW PIECES TOGETHER TIGHTLY so there is no gap between them. In order to create a strong joint, the screw needs to draw the two pieces together so that they firmly press against one another. This creates friction between the pieces, which increases the shear strength of the joint. A screw has much less shear strength than a nail of similar length, meaning that it is less able to resist shear forces—that is, opposing forces trying to slide the two pieces past each other. But screws have threads, which gives them grip strength, meaning that they are able to pull two pieces together tightly.

PREVENT WOOD FROM SPLITTING by predrilling pilot holes. I use a drill bit that snugly fits the shank of the screw (the part you'd be left with if you took off the threads).

Pilot holes aren't always necessary, though. If you are using screws that are #10 (about ³⁄₁₆ inch in diameter) or smaller and making only rough connections in lumber larger than 2×2, a pilot hole usually isn't needed—especially if the wood is still a little green and pliable.

If you don't predrill a pilot hole, often the two pieces you are screwing together will push apart as the screw starts to move into the second piece. When this happens, back the screw out of the second piece and then drive it back in. The hole that was started in the second piece will now act as a pilot hole. This makes a tight, strong joint.

AVOID DRIVING SCREWS IN WEAK SPOTS in the wood. For example, don't place screws too close to the edge of the wood. Leave at

least ½ to ¾ inch between the screw and the edge. In addition, avoid putting two screws on the same grain line, and don't place a screw anywhere that looks weak or is splitting already. Conversely, you should also avoid screwing into knots in the wood, which are usually too hard to screw through directly and may even break your screw.

SPREAD OUT THE SCREWS as much as possible. Spacing screws widely not only helps prevent splitting, it also helps the joint resist torsion (twisting force). If you fasten two pieces together with only a single screw, the pieces can spin around that screw. Adding a second screw locks the joint against spinning. The farther apart the two screws are, the more resistance to torsion the joint has. Additional screws pull more of the two pieces into strong contact with each other, resulting in more friction between the pieces, which helps resist both torsion and shear forces.

AVOID USING TOO MANY SCREWS, which can actually weaken wood. In addition, more screws add expense, both in materials and in time.

JOINTS WILL LOOSEN OVER TIME due to stresses on the joint and cyclical drying and absorption of moisture by the wood. Keep this in mind when designing screw joints: Don't rely entirely on the friction they create between two pieces to keep a joint tight, and retighten the screws if the joints do get loose over time.

And, of course, **corded drills** and **hand drills** still exist. Corded drills are lighter, more powerful, and less expensive than cordless drills, and they are a good option if you're always going to be near an outlet. Old-fashioned hand-crank drills are much lighter than cordless drills, never need to be charged, and don't make a lot of noise. I often use one for quickly drilling small holes, a task that doesn't take much energy.

Drill bits come in many styles. **Twist bits** are the most common type of bit for drilling small holes (typically anything up to ⅜ inch with a hand-held drill) and **spade bits** are common for larger holes. Today, most twist bits are also suitably hard for drilling in metal. Other types of drill bits, including Forstner bits and hole saws, each have their own special purposes. Forstner bits are more expensive but they do some things that spade and twist bits can't. For example, while both spade bits and Forstner bits cut flat-bottomed holes, Forstner bits produce less splintering when they come out the far side of a through hole. And unlike a twist bit or a spade bit, a Forstner bit can be used with a drill press to drill partial holes in the edges of wood, creating rounded grooves.

A **hole saw** is kind of a hybrid between a saw and a drill bit that saws a round hole through the material. Hole saws are good for cutting larger holes in any material that is an inch thick or less, and because of their fine teeth, they are the best option for cutting larger holes in thin plastic pipes to avoid cracking the pipe.

I use my cordless drill for driving screws as well as drilling holes. A magnetic drive guide makes switching bits fast and helps hold the screw steady while you set it, which is especially helpful if you're driving in an awkward spot.

Another option for driving screws is an impact driver, which looks similar to a drill but is more compact and works a little differently, making it better for driving screws than for

A magnetic drive guide is helpful when you're using a drill to drive screws. The outer sleeve of the guide can be extended over long screws to prevent them from tipping as they're being driven into the wood. The bit can be changed easily when it wears out.

drilling holes. An impact driver literally uses impact forces to twist the screw, hammering the shaft around in a circle instead of applying an even twisting force like a cordless drill does. Because impact drivers develop high forces in short bursts of many little impacts, you need to use bits and socket sets specifically designed for impact drivers. Otherwise they can shatter. One major advantage of impact drivers is that they drive screws really well without the tendency to strip the heads of the screws the way drills do.

Cutting

When you're cutting wood, it's useful to understand that wood is a composite material made up of fibers that run the length of the wood and are held together by resin, a kind of natural glue (for more on composites, see Materials Properties in Plain English, page 198). The fibers are referred to as the *grain* of the wood. Cutting perpendicular to wood grain, or "across the grain," is called *crosscutting*, while cutting parallel to the grain, or "with the grain," is called *ripping*.

Cutting wood is mostly done with a saw, and saws have different types of blades depending on whether they're made for crosscutting or ripping. Crosscut blades are designed to slice through the wood fibers, and rip blades are designed to chip away the resin along with little bits of the fibers. There are also combination blades that do a reasonably good job of both crosscutting and ripping, and those are generally the most useful on the farm.

When looking at saw blades, the other feature you'll notice is the number of teeth. Blades with fewer teeth that are spaced more widely apart will cut faster and leave a rougher surface, while blades with more teeth per inch (TPI) will cut more smoothly but will take a bit longer to get through the wood.

I regularly use four kinds of saws on the farm: a **hand saw**, a **circular saw**, a **corded jigsaw**, and a **table saw**. The hand saw I use the most is a basic 15-inch crosscut saw with 12 TPI. I mostly use it to cut small pieces and to make single cuts when I don't feel like pulling out the larger circular saw, which needs to be plugged in.

Of all my saws, my circular saw is the one I probably use the most. Circular saws are great for making straight cuts of all sorts, but especially crosscuts. They come in different sizes, and the most common size uses a 7¼-inch blade. This blade is big enough to cut through most common materials, even with the blade at an angle, but not too big to make the saw difficult to handle.

Combination blades for circular saws will crosscut and rip reasonably well, in addition to cutting through materials like plywood, which is made up of multiple layers of wood oriented in different grain directions. Because saws require a lot of power to run, I recommend avoiding a cordless saw as your primary saw, since it will be somewhat limited in the number of cuts it can make on a single charge and will cut more slowly.

Jigsaws have relatively narrow, short blades, and you can get blades with teeth designed for cutting many types of materials.

While jigsaws can be used to cut straight lines, they are particularly suited to making curved cuts and cutting thinner material. They're also ideal for cutting out notches and other small details.

A table saw is basically a circular saw with the blade fixed to a table. Instead of moving the blade across the wood, you move the wood across the blade. Table saws typically come with a rip fence and a miter gauge. The rip fence sits parallel to the blade and allows you to slide your board along it to make a very straight, accurate cut. The miter gauge slides in a groove on the table parallel to the blade. It can be set perpendicular to the blade for straight crosscuts or to any angle up to 45 degrees for making angled cuts. Table saws are generally more accurate and powerful than handheld saws, and they have larger blades, but they're also harder to move around and can be awkward to use when you're working with larger pieces of wood.

There are many other types of saws, but those four are good ones to start with. If you learn how to use them well, you can make pretty much any cut you'll need on the farm. When you're first starting out with power saws, learn how to use them safely. The larger the saw, the more potential it has to do serious damage. Many people have lost fingers or worse to power saws because they weren't using safe practices. But it is not difficult to use a power saw safely if you know how. See page 9 for some basic tips.

Shaping and Smoothing

When building most farm tools, you won't need to smooth anything to perfection. Use coarse 80-grit or 120-grit **sandpaper** wrapped around a wooden block to knock off the splinters from cut edges. In addition, a sharp ½- or ¾-inch **chisel** is useful to notch out corners if necessary.

Clamps

When cutting and assembling, it's often helpful to have an extra set of hands, or maybe even a few extra sets, to hold things in place. If you don't have a helper nearby, use 12-inch trigger **clamps** (I use the Quick-Grip clamps made by Irwin). These can be tightened easily with one hand while your other hand holds the work. Clamps are also good for holding glue joints tight while they dry.

Measuring and Marking

I use two types of measuring tapes on the farm: a 25-foot **retractable tape measure** for wood- and metalwork, and a 100- or 200-foot **reel tape** to mark out beds in the field, irrigation lines, and other longer distances. Retractable measuring tapes are usually made from spring metal and have a Mylar coating. They do wear out over time but are pretty durable if you don't abuse them. The little hook on the end of the tape pulls out very slightly when you catch it over the end of a board, and it pushes in slightly when you measure up against a surface. This clever design feature means the measurement is good to about $1/32$ of an inch whether you use it with the hook pushing against or pulling on a surface. The slight movement adjusts for the thickness of the hook.

Reel tapes are very flexible and can act as a straight line when under tension, but they don't provide accurate measurements if they are loose. In the field I usually fix the end of the reel tape to the ground with a 6-inch ground staple and pull against that. Be aware that reel tapes usually have one side in feet and inches and the opposite side in feet and tenths of feet (some surveyors and engineers work in tenths of feet, not inches).

In addition to measuring length, you will often need to check that two pieces are exactly perpendicular to each other at a corner, forming a perfect 90-degree angle; this means that they are "square." To check for square, use a 12- or 18-inch **framing square** or a **speed square** (the

standard size is 7×7 inches). A speed square can also be used to guide the foot of a circular saw or jigsaw to help you make straight 90-degree cuts (perpendicular to a board edge) or at an exact 45-degree angle.

If you have a large square or rectangular layout, and you want to check the entire layout for square (90-degree corners), first double-check that the parallel sides are the same length. Next, use a standard or reel tape measure to measure the diagonals—the distance between opposite corners; when the measurement is the same for both diagonals, the layout is square.

Another way to check for squareness, regardless of the project size, is to use the Pythagorean theorem, or the 3-4-5 triangle. (Remember back to geometry class?) A triangle with sides of 3, 4, and 5 units—whatever the units might be—has a perfect 90-degree angle between the sides that are 3 and 4 units long.

If you have two long, straight pieces of wood that are supposed to be at a 90-degree angle, you can mark one at 3 feet from the corner and the other at 4 feet. The diagonal between the 3-foot and 4-foot marks should be exactly 5 feet long. If the diagonal is more than 5 feet long, the angle is more than 90 degrees. If the diagonal is less than 5 feet long, the angle is less than 90 degrees. Either way, adjust the angle until you get the diagonal to be exactly 5 feet.

To know when something is exactly level or exactly vertical (plumb), use a **bubble level**. These are available in many lengths, from very short to very long, but on the farm you probably won't need anything much bigger than a 6-inch level, sometimes called a torpedo level. Occasionally a 2-foot or 4-foot level is useful, and a 4-foot level works well as a straightedge for marking cuts on plywood.

For most marking, a regular wooden #2 **pencil** works really well. To sharpen your pencil without a sharpener, use a small knife to slowly whittle away the tip as you rotate it. You can also use a knife to scribe a mark on wood if you need a fine line or if you don't have a pencil handy. One nice thing about marking a cut line in wood with a knife is that it precuts the surface grain, which reduces splintering. For really long straight lines (such as marking lines on the full 8-foot length of plywood), use a **chalk line**. This common tool consists of a box that holds powdered chalk and a reel of string that extends through a hole in the box and is tied to a small hook. As the string is reeled out, it gets coated in chalk. The string can be pulled tight against a surface and then snapped against the surface, leaving a straight line of chalk.

In many cases there are clever ways to measure without using a special measuring tool at all, and frequently it's more precise to skip the measuring tool. For example, the reason to make a table 30 inches high instead of 30³⁄₁₆ inches high is usually just because nice round numbers are easier to measure with a measuring tape. In other words, a lot of measurements are somewhat arbitrary.

But what *is* important is measuring accurately when two pieces need to match. In these situations, transferring the measurement directly from one object to another is often the fastest and most accurate method. Going back to the table example, simply cut one table leg and then use that leg as a template to mark the measurement

Pythagorean Theorem

As noted on the facing page, you can use the Pythagorean theorem to make sure an angle is square. This works with any multiple of 3, 4, and 5, so you can use longer measurements for larger pieces and smaller measurements for shorter pieces. Just multiply all three measurements by the same number. For example, you could multiply by 2 to get 6, 8, and 10, or multiply by 1.5 to get 4.5, 6, and 7.5. The longer the measurement, the more accurate the result.

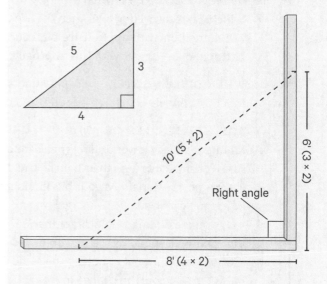

for the other three legs. Always use the same template piece for marking all of the pieces that need to match. This way, any small difference between the first and second pieces won't transfer to the third piece and multiply the difference.

Basic Metalworking Tools

THERE IS A LOT OF OVERLAP BETWEEN woodworking and metalworking tools. Here are some of the tools that I've added to my collection specifically for working with metal.

Drilling

You can often use the same **twist bits** to drill metal that you use to drill wood, although it is more important that the bit is sharp when drilling through metal. Drilling metal also requires more pressure and sometimes less speed. See page 16 for some tips on drilling metal.

TIPS DRILLING METAL

- **MARK THE CENTER OF THE HOLE WITH A CENTER PUNCH** before you start to drill into metal. The indentation from the punch will help keep the drill bit from wandering when it starts out.

- **DRILL A PILOT HOLE** when drilling with a twist bit larger than ¼ inch. Start by drilling a ³⁄₁₆-inch or ¼-inch pilot hole, and then widen the hole with the full-size bit. Your larger bits will work better and last longer with this approach.

- **WHEN DRILLING STEEL**, you will need a drill bit that is rated for cutting steel—HSS (high-speed steel) or better.

- **ADJUST THE SPEED OF YOUR DRILL BIT.** Harder materials require relatively lower drilling speeds and more downward pressure. For small bits (less than ¼ inch), the drilling speed will be similar for most materials. As your bits get larger, the rotational speed needs to be slowed.

  If you are drilling a hole larger than ⅝ inch, in harder materials the rotational speed of the bit can be less than one-quarter what it would be for a smaller hole in the same material. In softer materials such as wood, the change in rotational speed is less extreme. For example, if you drill a ¹⁄₁₆-inch hole in wood or steel, your bit should spin at about 3,100 revolutions per minute (rpm), but if you drill a ½-inch hole, the bit speed in wood might only be 1,100 rpm and in steel it could be 750 rpm or less. (Consult a drill press speed chart if you're using a drill press.)

  The drilling should generate decent-size "chips" of metal (larger than grains of sand). If it doesn't, you are spinning the bit too fast, you are not pushing down with enough pressure, or the bit is dull. All of these scenarios will result in heating up and dulling both the metal and the bit. If you are building up too much heat, you can use a cutting oil, but oil probably isn't essential if your bit is sharp.

Cutting

Depending on the type of metal and the thickness you're cutting, you'll usually need a different blade and probably a different saw than you use for cutting wood. And just as with drilling metal, you need to be careful that you use appropriate pressure and speed when cutting metal. Use a **hacksaw** or **jigsaw** with a metal-cutting blade to cut small objects such as steel bolts. To cut larger quantities of metal tubing, pipe, and flat stock, use a 10-inch abrasive **chop saw** or a metal **band saw**. The abrasive chop saw is loud and messy, creating big rooster tails of sparks as it cuts, but it makes neat, straight cuts, while the band saw is slower, quieter, and relatively cleaner.

You can also cut through metal using an **angle grinder** with an abrasive cutoff wheel. Like the chop saw, it is loud and messy. In addition, the combination of high rotational speeds and relatively fragile blades on this tool, along with the large quantities of sparks and dust it creates, means you should pay particular attention to good safety practices when using it for cutting.

Shaping and Smoothing

To shape and smooth metal edges, **hand files** work well. These are available in a huge array of sizes and shapes; the two I use the most on the farm are 6-inch and 12-inch single-cut mill bastard files. Single-cut files leave a smoother finish than double-cut files do. *Mill* refers to the shape, which is slightly tapered and flat. *Bastard* refers to the coarseness of the file and is somewhere between coarse and fine.

Use an **angle grinder** or **bench grinder** to prepare surfaces for welding or to take off materials faster than you could with a file. With the angle grinder, you hold the tool, while the piece you're grinding remains stationary. With the bench grinder, the tool is fixed to a workbench

and you bring the workpiece to it. Both types of grinders can be fitted with different kinds of wheels, including wire wheels and flap sanders, which are handy for taking off surface rust and loose paint.

Clamps

There are a couple clamps that are especially handy for metalwork. For holding pieces together for welding, **C-clamp locking pliers** are the most useful for the kinds of projects in this book. Regular **C-clamps** work well and provide a tight hold but usually require two hands to operate, so they are harder to use. Avoid plastic quick-grip clamps for welding because they don't hold up to the heat, although they can work well if they are used well away from hot surfaces.

Measuring and Marking

You can generally use the same tools for measuring metal that you use for measuring wood. Mark metal with a **Sharpie** or **soapstone pencil**. I rarely need a precise line when working with metal, but if I do, I first darken the area with a Sharpie and then use the tip of a pointy metal tool to scratch a line through the ink and into the metal. A **scriber** is the tool designed for this task, but a homemade **poker tool** can make a good scriber. A poker tool is also handy for digging dirt out of small spaces or just for poking things.

To make a poker tool, start with a 6-inch piece of wood (from a tomato stake or some similar scrap) to use as a handle and a 4- to 6-inch piece of straight steel wire (an old bicycle spoke, or even a coat hanger, works well for this). Using a bit that is just a hair smaller than the wire, drill a hole 1–2 inches deep into the end of the handle. Push the wire into the hole using a pair of pliers. Sharpen the end of the wire to a point with a grinder.

Workbenches and Toolboxes

IT'S NOT ESSENTIAL to have a **workbench** or a **toolbox**, but I find them handy. For a temporary workbench, you can simply lay a few boards across a pair of sawhorses. You may also decide you like having a dedicated workbench that you don't mind scratching up or getting odd bits of glue and paint on, and which you can replace when it wears out. My workbench is essentially a version of the Potting Bench (page 40), without the mixing tub, but with freestanding legs and a thicker plywood top.

When it comes to storing tools, what's important to me is that the tools are organized, easily accessible when and where I need them, and kept dry and out of the sun to prevent them from wearing out prematurely. Mostly I use wooden shelves with designated locations for different tools. I also have a couple of very simple toolboxes that I use when I'm taking my tools somewhere outside my workshop. The toolbox I like the best is one I slapped together from old 1×6 fence boards and a few screws more than 20 years ago. It has a handle in the middle and two long compartments for tools on either side.

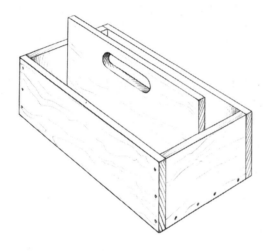

Years ago I found a design for a three-legged sawhorse made from 2×4 and 2×6 boards. The design has a broad, slotted top that allows a piece of 2× wood to be clamped in the slot, or a saw blade to cut through a piece of wood freely while the wood is supported on both sides. On uneven ground, the three-legged design is fundamentally more stable than four-legged designs. I often use this sawhorse as a workbench for small projects.

Treating Your Tools Right

KNOWING HOW TO USE your tools properly—and how not to use them—are important steps in giving them a long and useful life. For example, it's inevitable that you'll sometimes use a wrench to tap on things. But acknowledging that a wrench is not a proper hammer—and that hammering with a wrench is potentially harmful to the wrench, if not dangerous to you—will remind you that if you really need a hammer, you should go get one, and not use the wrench that's in your hand just because it's there.

To maintain your tools' usefulness and prevent them from wearing prematurely, keep them dry and store them appropriately. If metal tools get wet, dry and oil them before storing them in a dry place. Similarly, dirty tools will wear from the grit of the soil. The soil itself can also hold moisture and other elements that cause corrosion. To repel water and prevent corrosion on the metal parts of the tool, apply a thin coat of machine oil.

Wood handles also need to be stored dry and oiled to prevent both wet rot and corrosion of their natural resins. On wood handles, use a relatively heavy oil, such as linseed oil, which will penetrate the pores in the wood and dry to a hard finish. This helps protect the wood from wear and water. Both wood and plastic handles will break down with sun exposure, so they are best stored out of direct sunlight.

All blades cut best when they are sharp. This allows them to cut straighter and with less force, meaning they are easier to control and less likely to accidentally go somewhere unintended. Many saw blades and drill bits aren't easily sharpened without specialized tools, but basic blades such as knives, chisels, hoes, and even pruners are relatively easy to sharpen once you know how. To sharpen carbon steel blades, such as those on most hoes, I recommend using a mill bastard file, as described on page 17. To sharpen harder stainless and spring steel blades, such as those on most knives and stirrup hoes, I recommend using a medium pocket stone with water.

I try to sharpen all of my blades regularly, before they become completely dull and need major work to get them back into shape. I sharpen my most commonly used knives, hoes, and chisels daily before use. Resharpening a blade that is only slightly dull is quick and easy and it makes them much, much easier to use.

A basic blade angles to a point. The angles are called the bevels, and they may be on one or both sides of the tip. Most of the wear on the cutting edge of a blade is at the tip, which is the weakest part. The edge becomes rounded and needs to be resharpened. To resharpen the tip, take material from the bevels while maintaining the original angle, until enough material has been removed that the edge is restored.

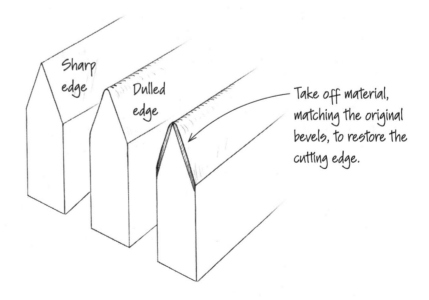

Sharp edge

Dulled edge

Take off material, matching the original bevels, to restore the cutting edge.

Basic Materials

MANY OF THE MATERIALS I use in the designs in this book were chosen as much out of convenience as a result of an in-depth engineering analysis. What follows here is a brief overview of the basic materials used in many of the designs and why I think they're good choices. There are many other materials that might work well for you when modifying these designs or designing your own tools. Some of these other materials may be readily available for you, and some have superior properties to the materials I've used. For more detail on the technical differences between materials types, see Materials Properties in Plain English on page 198. In that section I also explain what information I consider when choosing materials for a project.

Wood

Wood is one of the materials I work with most often. It's commonly available and inexpensive, and woodworking tools are also commonly available and relatively inexpensive. The most common grades of wood that I use on the farm are construction grade or "standard or better" SPF, which stands for spruce, pine, or fir. (Where I live in the Pacific Northwest, Douglas fir is common, so nearly all of the wood in the versions I built to test these designs is Douglas fir.) I also use a lot of CDX plywood, which is one of the most common types of plywood. Its C- and D-grade surfaces are a bit rough, but very serviceable.

For most of the wood tools I build, I use untreated wood and leave it untreated and unpainted. Paint and other surface treatments, if not properly maintained, can cause contamination. This is especially a concern if any of the surfaces come into regular contact with food. Paint and other surface treatments can also hold moisture and accelerate rot, instead of helping to preserve the wood. The natural rot resistance of Douglas fir and many other woods allows them to last a long time—well over a decade in my experience—even when they get wet repeatedly. This varies by wood species, climate, and use.

Steel

The steel I use for the Hand Cart (page 64) is hot rolled low-carbon steel. You won't find this kind of steel at most hardware stores or even big box outlets, but it's very commonly available from metal suppliers and at scrapyards. It's probably the most common steel out there and is one of the easiest to weld.

Other projects in the book call for black pipe or galvanized pipe. These are available at most hardware stores, along with many types of fittings. Black pipe can be welded safely, but galvanized metal produces toxic fumes when heated to welding temperature. The difference between black pipe and galvanized pipe is the type of coating used to protect the pipe from corrosion. Galvanized pipe is more resistant to corrosion, but in practice I find black pipe is plenty corrosion resistant for all of my tool designs. Black pipe is also less expensive and has a smoother surface, which is especially important for parts that are handled regularly.

Plastic Pipe

Most of the projects in this book that call for plastic pipe are for use with irrigation (see Chapter 4). For water distribution I almost always use flexible, black polyethylene tubing that I get from irrigation suppliers. It is available in nominal sizes such as ½ inch and ¾ inch, though frequently the actual sizes vary slightly from supplier to supplier. I recommend getting your fittings from the same supplier you get your tubing from to ensure a good fit.

For small-farm and garden building projects, polyvinyl chloride (PVC) pipe is a ubiquitous material due to its low cost, light weight, and easy workability. The two basic types of PVC pipe are schedule 40, which is white, and schedule 80, which is gray. Schedule 80 is thicker and stronger, but it is also much more flexible. The additives in schedule 80 make it more resistant to degrading from sunlight exposure. If your PVC pipe will be exposed to direct sunlight, painting it will make it last much longer.

There aren't any projects in the book that use greenhouse film, but it's worth noting that when PVC is in contact with greenhouse film (as I commonly see on farms), the film will degrade prematurely due to the pipe's off-gassing of chlorine. Painting the pipe with water-based paint helps reduce this problem.

PVC manufacturing and disposal releases toxic chemicals, so I try to avoid using PVC when I can find a good alternative. ABS pipe, which is black, is similar to PVC, and in many cases it works as well as or better than PVC, with a slightly less toxic environmental profile. Most hardware stores commonly carry both of these pipe materials, although they may not carry all of the same sizes and fittings for both. Each kind of pipe requires a special cement for bonding, and the cements are not interchangeable; PVC requires primer with the cement, while ABS does not. In addition, unprotected ABS is more prone to warping in sunlight than PVC is, so it's a good idea to paint ABS if it is going to be exposed.

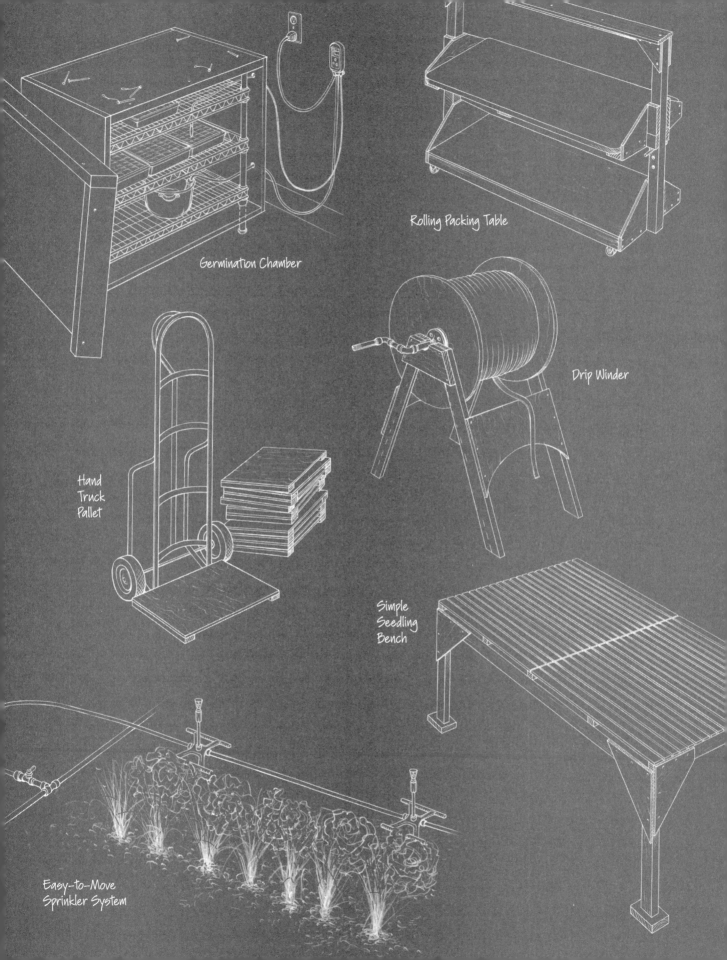

Germination Chamber

Rolling Packing Table

Drip Winder

Hand Truck Pallet

Simple Seedling Bench

Easy-to-Move Sprinkler System

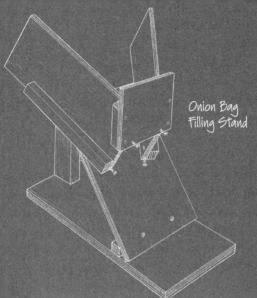
Onion Bag Filling Stand

CSA Boxes

PROJECTS

Mini Barrel Washer
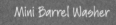

Hand Cart

Hoop Bender

CHAPTER 2
GREENHOUSE TOOLS

YOU'LL FIND THAT THE WORD *TOOLS* IS USED LOOSELY throughout the book. The three tools in this chapter are also sometimes called *furniture*. They are basic pieces you can build for a greenhouse to make it a better place to produce healthy seedlings for your farm or garden. These three pieces of greenhouse furniture are tools I've used weekly for years now to help me grow all of the seedlings for my own little farm. Together, they make a complete system that streamlines my workflow and allows me to work safely and efficiently. They also help create the right conditions for growing consistent, strong seedlings. Read on for insights into the designs and instructions on building your own versions of these tools.

SIMPLE SEEDLING BENCH

This simple, lightweight yet sturdy bench is used for growing seedlings in plug trays in our greenhouse. We've been using benches like this one for decades and have made several design refinements along the way. It's built with basic off-the-shelf lumber in standard sizes. It's also easy to customize to get the height, width, and length just right for your needs.

In this example, the bench is 36 inches tall, 42 inches wide, and 8 feet long. I'm 6 feet tall, and I found 36 inches a convenient height for the bench, allowing me to comfortably pick up and set down the trays (my elbow height is about 41 inches, for reference). This height also means I don't need to bend over to get a good look at the plants. You can vary the height of your bench simply by changing the length of the legs.

DESIGN NOTES

The bench consists of a basic lumber frame reinforced with plywood at the corners. The frame is topped with evenly spaced wood strips, called lath, that measure about $\frac{1}{4} \times 1\frac{1}{4} \times 48$ inches. Lath is commonly sold in bundles of 50 strips. The gaps between the strips allow for airflow and water drainage and provide spaces for laying in tubing for bottom heat (see Using the Benches, page 31).

At 42 inches wide, the bench is designed to fit multiple rows of seedling trays that measure 10×20 inches (commonly called 1020 trays or "ten-twenties"). The trays are actually a bit longer than 20 inches, so the extra 2 inches of bench width means the trays don't hang off the edges, where they might get bumped. The 42-inch width also allows enough space for irrigation tubing down the center of the benchtop if you integrate sprinklers into the bench (again, see Using the Benches, page 31).

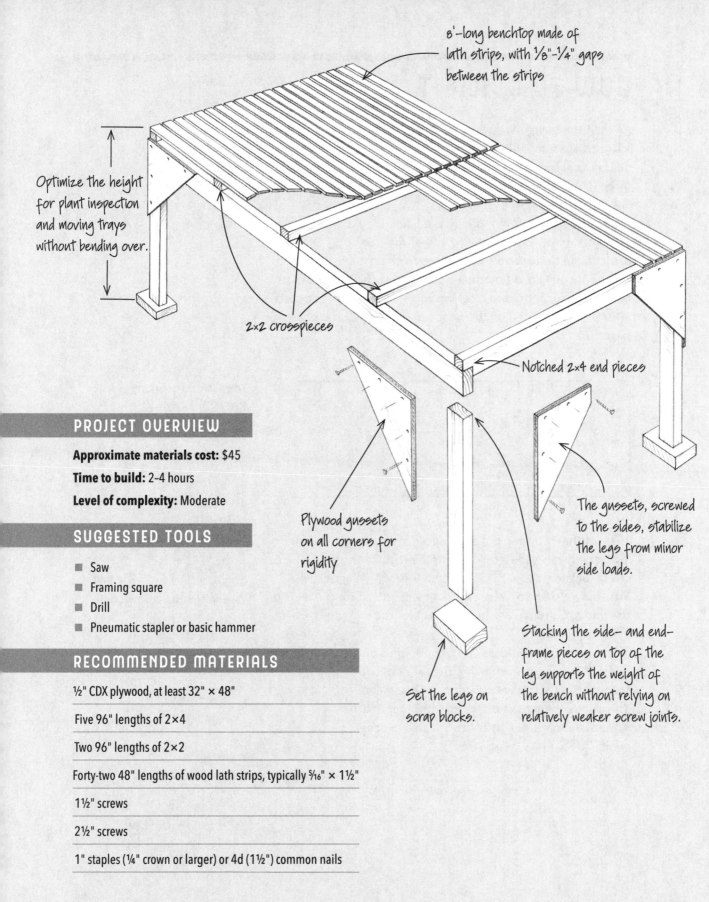

8'-long benchtop made of lath strips, with ⅛"-¼" gaps between the strips

Optimize the height for plant inspection and moving trays without bending over.

2×2 crosspieces

Notched 2×4 end pieces

Plywood gussets on all corners for rigidity

The gussets, screwed to the sides, stabilize the legs from minor side loads.

Set the legs on scrap blocks.

Stacking the side- and end-frame pieces on top of the leg supports the weight of the bench without relying on relatively weaker screw joints.

PROJECT OVERVIEW

Approximate materials cost: $45

Time to build: 2–4 hours

Level of complexity: Moderate

SUGGESTED TOOLS

- Saw
- Framing square
- Drill
- Pneumatic stapler or basic hammer

RECOMMENDED MATERIALS

½" CDX plywood, at least 32" × 48"

Five 96" lengths of 2×4

Two 96" lengths of 2×2

Forty-two 48" lengths of wood lath strips, typically ⁵⁄₁₆" × 1½"

1½" screws

2½" screws

1" staples (¼" crown or larger) or 4d (1½") common nails

HOW TO BUILD IT

1 **Cut the triangular gussets.** Mark two 16"-wide strips on the end of a full-width sheet of ½" plywood. Then, mark both edges of each strip at 16" intervals and join the marks to make three 16" squares. Draw diagonal lines between the marks to create two triangles per square.

Cut the strips from the sheet, then cut along the diagonals and cross marks to make the triangles. You need only two triangles from the second strip, for a total of eight. (It's helpful to trim about an inch off the sharp points of each triangle at this point to keep them from catching and snagging pants pockets after they're installed on the benches; see the illustration on page 30.)

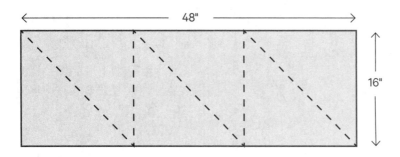

The pattern for cutting triangular gussets from a sheet of plywood

2 **Cut the 2× pieces.** Cut four pieces of 2×4 at 29½" (or the length you've determined) for the legs. Cut two 2×4 pieces at 96" for the sides of the frame (full 8-foot 2×4s are often a little longer than 96"). Cut two 2×4 pieces at 42" for the ends of the frame. Cut three 2×2 pieces at 42" for the crosspieces.

3 **Notch the end pieces.** Notch both ends of the 42" 2×4 end pieces so that they are left with stubs about 1½" thick and 1½" long on each end. This allows them to sit on top of the 2×4 frame sides at the same height as the 2×2 crosspieces (which are actually 1½" square).

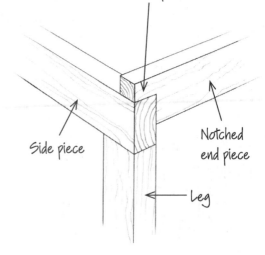

This stub should be the same height as your 2×2s and just long enough to be flush with the 2×4 side piece.

Side piece

Notched end piece

Leg

4 **Assemble the sides and legs.** Position one of the plywood gussets over one of the 2×4 side pieces so the square edges of the gusset are flush with the top and end of the side piece. Fasten the gusset in place with three 1½" screws along its top edge. Butt one of the legs against the bottom of the side piece, as shown, flush with the gusset. Check the leg with a framing square to make sure it is square to the side piece, then fasten the gusset to the leg with two screws. Repeat the same process to assemble the other side and leg pieces.

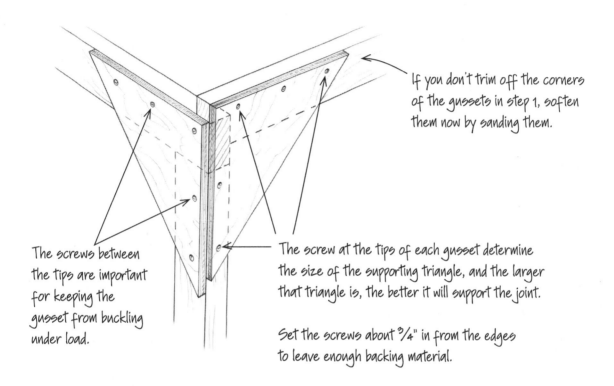

If you don't trim off the corners of the gussets in step 1, soften them now by sanding them.

The screws between the tips are important for keeping the gusset from buckling under load.

The screw at the tips of each gusset determine the size of the supporting triangle, and the larger that triangle is, the better it will support the joint.

Set the screws about ¾" in from the edges to leave enough backing material.

5 **Join the sides and ends.** Attach two gussets to each of the 2×4 end pieces, flush with the top edge and end of the 2×4 so that the plywood covers the notch. Stand up the two side-and-leg assemblies and fit them together with the two end assemblies. The notches should rest squarely on the top edges of the side pieces. Clamp the parts together at the corners (or have a helper hold them), check each corner with a framing square, and then fasten each gusset to the adjoining leg with two 1½" screws.

6 **Add the crosspieces.** Measuring from the outside edge of one of the end pieces, mark the top edge of each side piece at 24", 48", and 72". Position a 2×2 crosspiece across the bench frame so the crosspiece is centered on two opposing marks. Drill a pilot hole through each end of the 2×2 and fasten it to the side pieces with a single 2½" screw in each end. Install the remaining two crosspieces the same way.

7 **Install the lath.** Spread 21 pieces of lath evenly across the bench top so their ends are flush with the center of the middle crosspiece. The two outer slats should be flush with the ends of the crosspieces. Fasten the slats to the end pieces and crosspieces with a pneumatic stapler, if you have one, or use 4d common nails (the large heads on common nails help hold down the lath). Repeat the process at the other end of the bench. The slats on either end should match up at the center of the middle crosspiece.

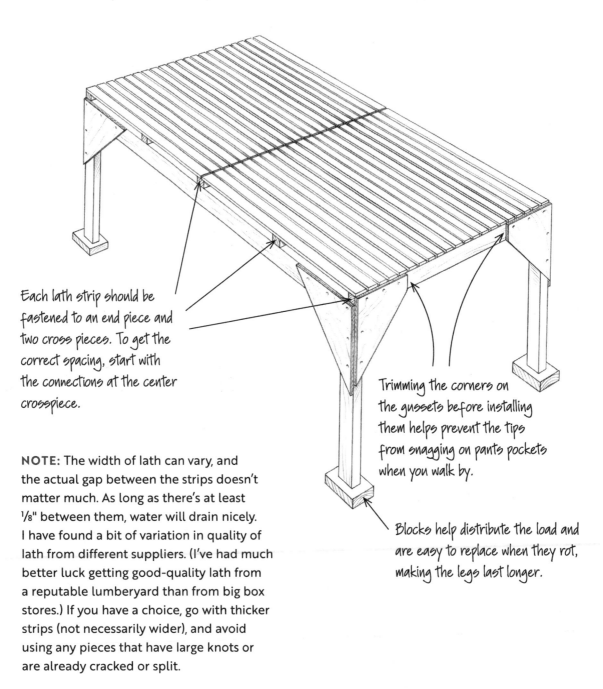

Each lath strip should be fastened to an end piece and two cross pieces. To get the correct spacing, start with the connections at the center crosspiece.

Trimming the corners on the gussets before installing them helps prevent the tips from snagging on pants pockets when you walk by.

Blocks help distribute the load and are easy to replace when they rot, making the legs last longer.

NOTE: The width of lath can vary, and the actual gap between the strips doesn't matter much. As long as there's at least $1/8$" between them, water will drain nicely. I have found a bit of variation in quality of lath from different suppliers. (I've had much better luck getting good-quality lath from a reputable lumberyard than from big box stores.) If you have a choice, go with thicker strips (not necessarily wider), and avoid using any pieces that have large knots or are already cracked or split.

USING THE BENCHES

At our farm, we use these benches for starting all of our seedlings. Occasionally in the late season we also use the benches for curing onions or squash. I have all of our benches set up with automatic irrigation to save time and to free us from being tied to the greenhouse every day of the week.

To avoid having dripping irrigation nozzles hanging above the benches, I integrate upright Ein-Dor sprinklers mounted on thin metal rods installed into the benchtops. Holes drilled partway into the center of the 2×4 end piece and 2×2 crosspieces are the perfect places to insert the metal rods and stand up the sprinklers. I then run the distribution tube for the sprinklers down the center of the benchtop and loop the spaghetti tube that feeds the sprinklers under one of the lath slats to hold the tubing in place.

I use sprinkler heads designed to throw water in a 4-foot radius and mount them every 2 feet (on the crosspieces) for more even coverage, which is more important to me than the wasted water from overspray. We set up every bench with its own valve so we can turn each bench's water supply on or off, and there is a hose connection between each set of benches that makes it easy to separate the benches and move them around if needed.

Between waterings and when there aren't any seedling trays on the benches, most of the wood dries out, so we haven't had any trouble with wood rot. The exception is the bottom of the bench legs, which tend to wick water from the ground and start to rot eventually. To protect the bench legs, I place each one on a scrap piece of 2×4 that is at

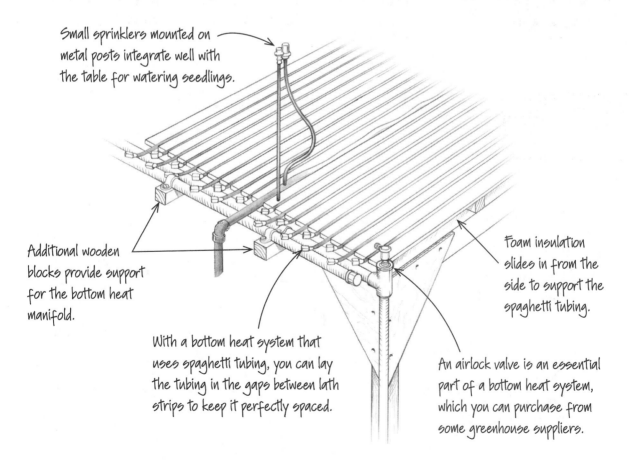

Small sprinklers mounted on metal posts integrate well with the table for watering seedlings.

Additional wooden blocks provide support for the bottom heat manifold.

With a bottom heat system that uses spaghetti tubing, you can lay the tubing in the gaps between lath strips to keep it perfectly spaced.

Foam insulation slides in from the side to support the spaghetti tubing.

An airlock valve is an essential part of a bottom heat system, which you can purchase from some greenhouse suppliers.

least 4 inches long. These sacrificial pieces will rot first, but they can be replaced much more easily than the bench legs. The scrap wood also helps distribute the pressure of the bench legs on the ground, keeping them more stable.

I've also used these benches with a spaghetti-tubing hot water system to provide bottom heat to seedling flats. The spaces between the lath hold the spaghetti tubing in place perfectly, without the need for plastic spacers. To run this kind of system, you need to add sheets of stiff foam board insulation just below the lath to support the spaghetti tubes and to hold in the heat; $\frac{1}{2}$-inch XPS foam board works well because of its stiffness and durability, but polyiso (or, technically, polyisocyanurate) is also a good option and has the added advantage of a foil face that helps reflect heat back up toward the roots. EPS foam board is inexpensive and also has a foil face, but I've found it a bit too delicate to work well for this application.

The foam board pieces should span the width of the table and fit between the crosspieces, making them approximately 42 inches wide and either 22½ inches or 21¾ inches long, depending on which section you're filling. To support the foam board, you can attach wooden runners made of lath strips or 1× lumber to the sides of the crosspieces to create a little lip for the edges of the insulation to sit on. This is easiest to do before you attach the lath for the benchtop, but I've done it after the fact as well. Leave just a small gap between the bottom of the benchtop lath and the top of the insulation, no more than ⅛ inch, to allow water to run off.

I've found it's best to angle this gap a bit so that the insulation has a slight slope, which will encourage water to drain to one side of the bench before it pools in the middle of the foam board and causes sagging. This also allows you to slide the foam board in and out from the side to clean it, as it inevitably collects debris.

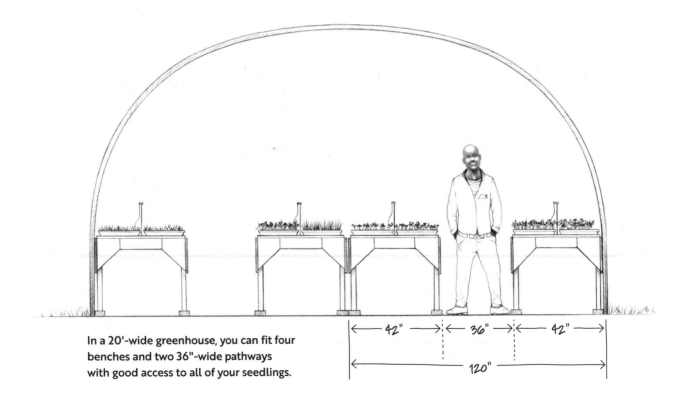

In a 20'-wide greenhouse, you can fit four benches and two 36"-wide pathways with good access to all of your seedlings.

42" 36" 42" 120"

Homemade Hoes

Three Versions of the Ubiquitous Farm Tool

The hoe is one of the most basic field tools used around the world. It's essentially just a blade on the end of a handle. Usually the blade is metal and the handle is wood, but there seem to be infinite variations. Over the years I've seen several handmade versions of hoes that are inexpensive to make and work just as well as hoes you might buy from fancy farm tool retailers.

Dr. Eric B. Brennan, a research horticulturalist at the US Department of Agriculture's research center in Salinas, California, has a YouTube video showing how he makes super-simple hoes for weeding. He builds them using only a bit of scrap metal strapping for a blade, a long reed (or bamboo) handle, and a piece of old bicycle tube, wrapped tightly, to tie it all together.

I've seen similar hoes that use scrap spring wire left over from building greenhouses (commonly called "wiggle wire") for the blade, attached to the end of an old broom handle. You can spend a good amount of money buying fancy versions of these wire weeder hoes, but they don't necessarily work any better than one you customize to your own needs.

A farmer in Northern California told me the Hmong farmworkers at his farm make their hoes from a bit of plumbing black pipe. This version is a little more involved than either Eric's hoe or the wire variation, but its construction is still very straightforward. After building a very hot fire, they heat the pipe, then hammer one end of the glowing hot steel into a fan-shaped blade. At the opposite end they leave some of the pipe shape intact and insert a broomstick handle into it.

These are only three examples of simple hoes, customized for the specific needs of the farmers who made them and built from materials they had on hand. Without a doubt there are countless other variations in use across the globe.

GERMINATION CHAMBER

The Germination Chamber is a box made of insulation board assembled around a set of wire shelves. This is a smaller, simpler-to-build, and easily customizable version of germination chambers I've seen on other farms and in building plans online.

In my version, each shelf fits three of our seedling trays side by side, with a little overhang on each end of the tray. There are four shelves, with enough space between each shelf to stack three trays. The pot of water that heats the chamber takes up some room, so at full capacity this Germination Chamber can hold 27 seedling trays. The whole unit fits under our seeding bench so it doesn't take up extra space in the greenhouse.

NOTES ON MATERIALS AND CONSTRUCTION

To build my chamber, I bought a used set of short Metro racks, the ubiquitous chrome-wire shelves found in restaurant kitchens. They are great for airflow and are relatively corrosion-resistant. I have friends who have made their own shelves with materials such as electrical metallic tubing (EMT conduit) or baker's sheet pan racks. You might try some of these materials (or others) depending on their availability and price, as well as the size of your seedling trays.

The insulated box is made with ½-inch-thick rigid foam insulation board on the sides, back, and bottom as well as the door. Most of this board has a heat-reflective facing (a common option) that faces into the box to help retain heat.

The top panel of the box is made with 1-inch-thick insulation board, adding a little strength to the assembly. The top panel is supported simply by the top of the shelving, and the rest of the box hangs from the top panel, with the bottom panel hanging about 1½ inches above the floor.

All of the box parts are taped together with insulation board tape (there is no additional support). You can use almost any type of insulation board as long as it's exterior-rated so it can handle a lot of moisture.

We heat our germination chamber with an electric 500-watt submersible aquarium heater set into a stock pot full of water (see Using the Germination Chamber on page 38 for details). If you live in a colder climate, you may need a larger heater and water container.

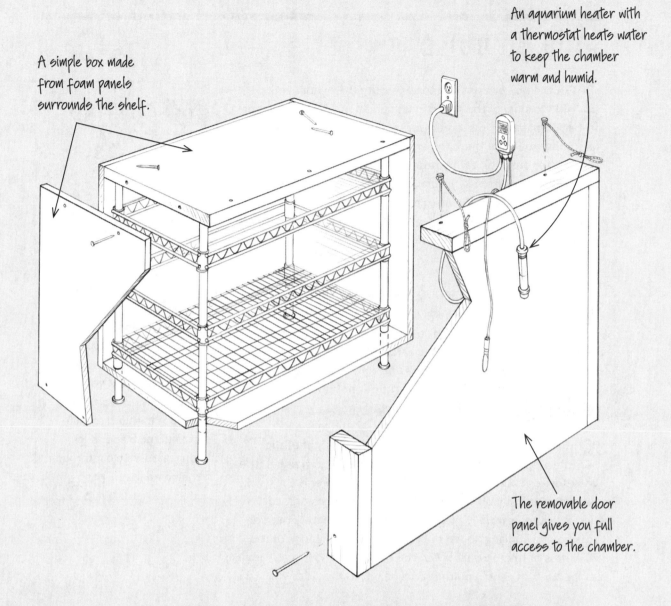

A simple box made from foam panels surrounds the shelf.

An aquarium heater with a thermostat heats water to keep the chamber warm and humid.

The removable door panel gives you full access to the chamber.

PROJECT OVERVIEW

Approximate materials cost: $60–$100

Time to build: 2–4 hours

Level of complexity: Simple

SUGGESTED TOOLS

- Utility knife
- Long metal straightedge

RECOMMENDED MATERIALS

½"-thick EPS rigid foam insulation board with heat-reflective film facing

1"-thick XPS or polyiso rigid foam insulation board

Foam board insulation tape or house wrap tape

Four 16d (2½") nails

Nylon cord (optional)

HOW TO BUILD IT

1 Cut the box bottom and top. Measure the overall width (side to side) of the shelving unit and add ¼". For the front-to-back dimension, measure your seedling trays and add 3" (assuming you want them to hang off the front and back of the shelves like ours do). Using these dimensions, cut the bottom panel of the box from ½" insulation board, and cut the top panel from the stiffer 1" insulation board. To cut the insulation cleanly, use a sharp utility knife, with the blade running along a metal straightedge. Take the time to make straight, clean cuts with square edges.

2 Make holes for the shelving legs. Place the bottom panel on the floor so the reflective facing is faceup, then carefully set the shelf unit on top so it is centered side to side and front to back on the board. Trace around the legs on the board, or simply press down on the shelf unit to create indentations in the foam. Remove the shelf and cut out holes for the legs all the way through the board with the utility knife.

3 Cut the box sides and back. Measure the height of the shelving unit. Measure the depth of the bottom panel (the back-to-front dimension) and add ½" (or whatever is the thickness of your insulation board). Using these dimensions, cut the two side panels from ½" insulation board. (Ultimately the bottom panel will hang above the ground by the distance of the thickness of the top panel). Cut the back panel so it is as wide as the bottom panel and the same height as the sides.

4 Assemble the top, sides, and back. Using insulation board tape or house wrap tape, connect the top panel to the back, with the back panel covering the thickness of the top panel. (You may find it easiest to tack the panels in place with short pieces of tape before attempting to fully seal the edges with long pieces.) Run tape along the full length of the outside and inside edges, being careful to keep the two pieces square to each other, the edges flush, and the tape free of wrinkles. Fit the side panels over the ends so the edges are flush with both the top and back. Again, fully tape the inside and outside edges. Any reflective facing on all panels should face inside the box.

TIP CUTTING FOAM BOARD

I find it easiest to make an initial shallow cut and then gradually deepen that cut with successive passes of the blade until I'm all the way through the insulation.

Some people like to use a snap-blade type of utility knife—the kind with a long blade that you can break off in sections. If you use this kind of knife, extend the blade so it's long enough to cut through the board in one pass.

5 **Add the bottom panel.** Set the shelving unit over the bottom panel, so the shelving legs fit into the holes. Place the taped-together panel assembly on the shelf, then lift the bottom so that it fits inside the back and side panels, tacking it in place with short lengths of tape. Check that all edges are flush on the outside, and then tape the bottom to the sides and back panel, as before, sealing all seams inside and out. Use the same tape to cover the exposed cut edges of the insulation board.

6 **Create the door.** The finished door will fit over the insulated box like the lid on a shoebox. All of the door parts are cut from $\frac{1}{2}$" insulation board. Measure the width and height of the front of the box. Cut the door panel to this size—but just a hair wider, so the door will fit snugly but not tightly over the front. Then cut three $2\frac{1}{2}$"-wide strips of insulation board, two for the sides of the door and one for the top. The top strip will overlap the

side strips, so it should be as long as the door plus the width of those two strips.

Tack the strips in place over the side and top edges of the door panel, sealing all seams inside and out and covering all cut edges with tape, as before. The bottom of the door does not get a strip of insulation board. This slight opening at the bottom of the door keeps water from collecting and makes the door easier to pull on and off. Because the opening is at the bottom and heat rises, heat loss is minimal.

7 **Secure the door.** Fit the door over the front of the box and hold it in place. Push two evenly spaced 16d nails through the top strip and into the top of the box. Do the same with one nail on each side, roughly 5" from the bottom edge. The nails serve as removable pins to secure the door. To prevent losing the nails and to make them easier to remove, tie short lengths of nylon cord or other type of string to them.

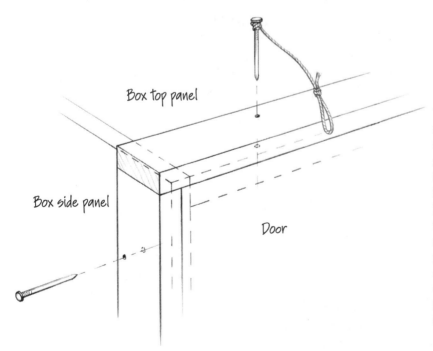

Box top panel

Box side panel

Door

Tie a short length of nylon cord (or any other type of string you have handy) to the nails to make them easier to remove. Taping the cord to the box with insulation board tape keeps the nails from disappearing—at least until the tape fails, at which point we've found it just as easy to place the nails on a workbench while we remove the door panel to open the chamber.

USING THE GERMINATION CHAMBER

The ideal temperature for germinating seedlings is around 70°F, depending on the seed type. In our climate, we don't need a lot of power to heat a small, well-insulated enclosure. For $25, I found a fully submersible 500-watt aquarium heater that came with a thermostat. I could also imagine using a smaller coil heater like the ones found in hardware stores for heating a mug of water.

The heater is plugged into a thermostat for electric greenhouse heat mats, and the thermostat plugs into the wall. The thermostat has a temperature probe on a wire that is hung inside the box and a readout unit on the outside. When the temperature inside the germination chamber falls below the thermostat's set point, it turns on the heater until the temperature rises a few degrees above that point.

For the water container, we use a stainless steel stockpot that sits on the bottom shelf.

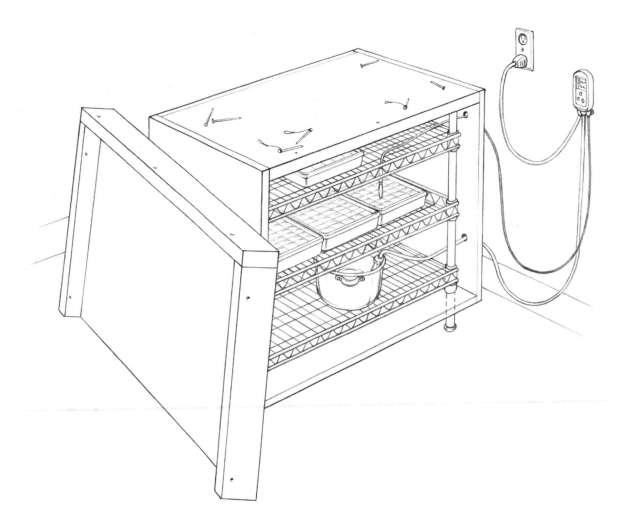

DESIGN NOTES

On the whole, this Germination Chamber has exceeded my expectations, both in how well it germinates seeds and in how long it has held up. After three years, the holes that the nails go through are getting a little loose, and the bottom panel has warped a bit. Other than that, the foam box is in good shape.

The biggest failure related to this Germination Chamber came when I forgot to fill the water tub for a week while the heater was on. All of the water evaporated, and the heating element burned a hole through the plastic tub. I switched to an old stainless steel stockpot and have paid a lot more attention to topping off the water every couple of days.

In the wake of that failure, a lot of folks have recommended using oil instead of water or installing a float valve like the ones used on stock tanks so that the container would fill automatically. While oil doesn't evaporate, it is more expensive than water, and I'm not sure how it might affect my submersible heater. Some people have also recommended a slow cooker filled with water or oil, or a light bulb as an alternative heat source.

The chrome wire shelves have become quite rusty in the warm, humid environment, but the rust hasn't made the chamber any less effective. Along the outside edges of the shelves, there is a bit of a lip that snags the bottoms of plug trays and makes loading a little tricky. To prevent that problem, I use a smooth-bottomed tray under the plug trays when loading them.

POTTING BENCH WITH MIXING TUB

In our greenhouse, we start the seeding process by blending purchased potting mix with a little balanced fertilizer and water to bring it up to the correct moisture level.

Once the blended potting mix is ready, we use it to fill our seedling trays. This potting bench facilitates both operations. It has a removable tub that holds a bagful of potting mix (we buy our mix in 2-cubic-foot bags) and sits at a comfortable height for mixing in the fertilizer and water. The tub is flanked by two benchtop work surfaces, sort of like the countertop areas next to a kitchen sink.

We set down the trays on one of the benchtops and scoop the potting mix out of the tub to fill the trays. The top of the tub is flush with the benchtop, so it's easy to sweep excess mix back into the tub after filling the trays. The bench has thin plywood along the back and one end to serve as a backsplash (another kitchen-like feature) to help contain the potting soil.

We also use a large metal bladelike painter's tool called a trim guide both to scoop the potting mix and to scrape the excess back into the tub. Any broad flat-edge tool would work for this, or you can make one from a board or thin plywood. Finally, an extra piece of plywood acts as a lid for the tub and prevents the potting mix from drying out if we don't use all of it at once.

NOTES ON MATERIALS AND CONSTRUCTION

Our potting bench uses a 24 × 36-inch concrete mixing tub, which has a capacity of approximately 21 gallons. The size and shape of your tub may vary, so it's best to build your bench based on your tub and make any modifications as desired. For example, you can build the bench as long, short, or deep as you want, and you can mount the bench at whatever height is most comfortable.

CDX plywood has one relatively smooth side and one rough side, although sometimes it's hard to tell the difference. If you want a really smooth surface for your potting bench, you can substitute MDO plywood for the CDX.

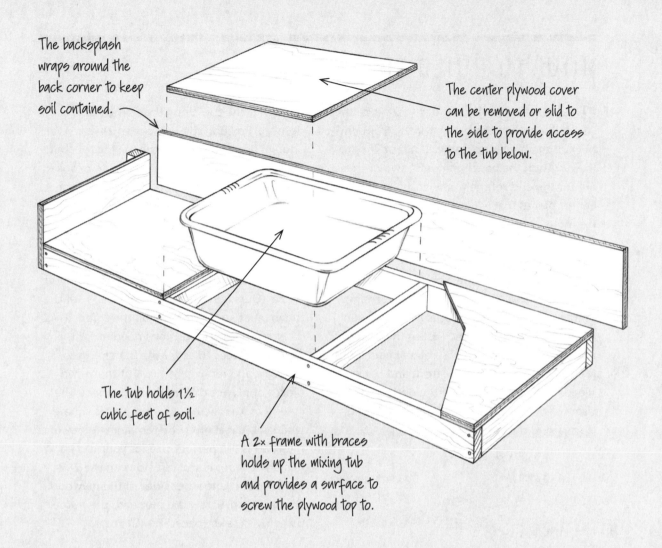

The backsplash wraps around the back corner to keep soil contained.

The center plywood cover can be removed or slid to the side to provide access to the tub below.

The tub holds 1½ cubic feet of soil.

A 2x frame with braces holds up the mixing tub and provides a surface to screw the plywood top to.

PROJECT OVERVIEW

Approximate materials cost: $65
Time to build: 2-4 hours
Level of complexity: Moderate

SUGGESTED TOOLS

- Tape measure
- Saw
- Drill
- Framing square

RECOMMENDED MATERIALS

Mixing tub
2×4 lumber
½" CDX plywood
¼" CDX plywood
One scrap of 2×2 lumber, as long as your backsplash is high
3" screws
1¼" screws

HOW TO BUILD IT

1 **Cut the frame pieces.** Cut two 2×4s to the full length of the bench for the front and back of the benchtop frame. Cut four 2×4s to the depth of the bench minus 3"; two of these are for the ends of the frame and two are center pieces that will support the tub. As an example, our bench is 90" long and 24" deep, so our end and center pieces are 21" long.

2 **Assemble the frame.** Set the front, back, and end frame pieces on edge, forming a rectangle with the end pieces fit in between the front and back pieces. Fasten the pieces at the corners with two 3" screws at each joint, screwing through the front and back pieces and into the ends of the end pieces. Use a framing square to make sure the frame is square at the corners.

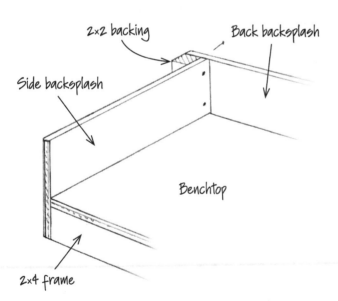

A scrap piece of 2×2 in the corner provides backing—a surface to secure the screws. I set the 2×2 on the outside of the corner and cut the back backsplash a little longer to accommodate it.

Center the tub along the frame (or as desired). Position the two center 2×4s just far enough apart to support the mixing tub, but leaving half of the width of each 2×4 exposed so that it can also support the plywood top in step 3. Fasten the center pieces with two 3" screws at each end.

3 **Cut and install the plywood tops.** Cut a piece of ½" plywood the same size as your frame. (Ours is 90" long and 24" deep.) Set the tub in place so its rim rests on the center 2×4s.

Measure the dimensions needed for the plywood pieces to cover the bench top to the right and left of the tub. Cut these two pieces of plywood from the large piece. (Be careful to keep your plywood pieces square when you cut them.) The remaining piece of plywood is the perfect size for your tub lid. Position the right and left sides on the 2×4 frame so the smoother sides of the plywood are facing up. Screw the pieces to the frame using 1¼" screws spaced 8"–12" apart.

4 **Add the backsplash.** Cut pieces of ¼" plywood (or use the ½" plywood, if you have some leftover) as a backsplash for the back and one end of the bench. The backsplash should extend above the benchtop at least 4"–6", but you can make it higher, as desired. (Our backsplash extends about 7" above the benchtop.) To keep the corner between the back and side backsplashes tight, you'll need to screw into some sort of backing; I used a scrap of 2×2. Attach the backsplash pieces to the bench frame using 1¼" screws. Stagger the screws on the 2×4s, placing every other one near the top edge and the others near the bottom edge.

LOOKING CLOSER ANGLED BRACING

Angled bracing, or diagonal bracing, is an incredibly common design feature because it helps joints between nonparallel pieces resist torque, or rotational forces (page 193).

The benchtop here makes a joint with the wall, with screws driven through its frame and into the wall. The top screw, about 3" above the frame's bottom edge, does most of the work to resist the rotational forces created by any weight on the benchtop. Without the angled brace, the benchtop would want to hinge at its bottom edge, where it meets the wall. If you put any downward force on the outer edge of the benchtop, that could be 24" from the axis of rotation. The resulting torque is 24" × the weight. To support that weight, the top screw has to generate an equal torque in the opposite direction, but it is only 3" from the axis of rotation, so the force on the screw is 8 times the weight (24 ÷ 3 = 8).

Adding an angled brace moves the axis of rotation down the wall, making the distances between the downward force and the rotation point and between the upper screws and the rotation point very similar. As a result, the forces are more similar. The larger the triangle made by the brace, the less force is needed to resist the torque, putting less force on the joint.

You can also calculate the compressive force that is put on the angled brace by using the sine of the angle between the brace and the benchtop. In the case of a 45 degree brace, it will be 1.4 times the magnitude of the downward force on the brace (1/sin45 = 1.4).

The compressive force on the angled brace is not as low as it would be if you used a vertical brace, but it is much, much lower than 8x from the example above. You can change the angle of the brace to further reduce the compressive force, but the increased material and space necessary for the brace may not be worth the amount by which you would reduce the forces on it.

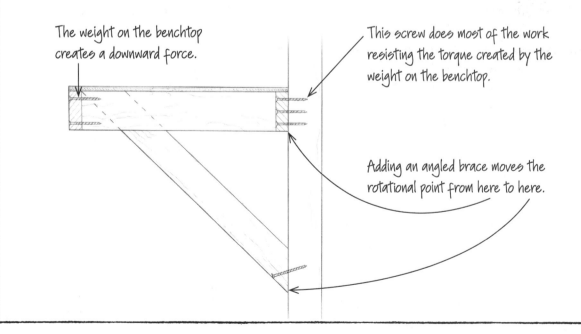

The weight on the benchtop creates a downward force.

This screw does most of the work resisting the torque created by the weight on the benchtop.

Adding an angled brace moves the rotational point from here to here.

DESIGN NOTES

Our potting bench is attached to the end wall of our greenhouse. The wall there is built with 4×4 posts, sunk into the ground, and 2×4 cross bracing. It is quite stable, but the bare ground in the greenhouse is not, so instead of building a potting bench with legs, I attached ours to the existing vertical posts in that wall.

There are advantages and disadvantages to this type of cantilevered construction. Among the advantages to cantilevering are that it takes a bit less lumber and there is more open space below the bench, since there are no legs coming down to the ground. The disadvantage is that it puts more force on the supporting parts and ultimately the end wall carries more load. This makes it even more important to create strong joints that won't fail.

With a cantilevered bench, the connection between the benchtop and the wall is essentially a hinge. The most effective way to prevent the benchtop from rotating around that hinge is to put diagonal braces from the front edge to the wall. I put my braces under the benchtop, which puts the braces in compression and keeps them free of the work area. It also puts the connection between the wall and the benchtop into tension.

Alternatively, you could build the braces so they angle upward from the benchtop. In that case, the braces would be held in tension by the weight of the benchtop, and the connection of the benchtop to the wall would be in compression. Chains or cables

are sometimes used for this kind of support, or you can use wood, which is strong when the tension is orientated with the grain of the wood.

Another result of this construction method is that you can't move the bench without partially disassembling it. You could make a freestanding version of this project using the same leg design used for the Simple Seedling Bench (page 26). With a freestanding bench, any weight placed on the benchtop is transferred directly through the legs to the ground, which is a very strong configuration. As long as the legs are perfectly vertical, the benchtop essentially will balance on the legs, even with a very weak joint between the top and the legs. You would still need to brace the legs against side forces, though, because people (and objects) lean on benches, and the legs aren't always 100 percent vertical when that happens.

For more information on forces generated in bracing and the basics of designing bracing, see Mechanical Principles in Plain English on page 191.

5 Mount the bench. Our bench is mounted to our greenhouse's end wall, which is framed with wood. The following directions explain how we mounted the bench, but there are many other mounting options, some of which are in the Design Notes on the facing page. Anchor the back 2×4 of the bench frame to the wall studs/posts at the desired height, using multiple 3" screws, while providing support along its front edge and making sure the bench is level.

6 Brace the bench front. To keep the front of the bench from sagging you need to support it in at least two locations near, or at, the ends. If the bench fits into a wall corner, you can anchor the front edge of the 2×4 at the end of the frame to the adjacent wall. This will create one of the bracing points.

At the free end of the bench, hold a piece of 2×4 at a 45-degree angle, starting at the front of the bench and angling down to the nearest wall stud/post. Mark a line where the board crosses the bench frame and the stud/post, then cut the board along the lines to create a brace. Screw the brace in place at both ends with 3" screws. Install another brace at the other end of the bench if your bench doesn't fit into a corner.

USING THE POTTING BENCH

We added this potting bench to our greenhouse a couple of years ago, and it works even better than I had hoped. Originally I planned to remove the plywood tub lid altogether when we used the bench and tub. Instead, I found that sliding the lid into the corner leaves it slightly overhanging the tub. This makes the lid a great work surface for filling trays because it's easy to scrape excess potting soil back into the tub.

I hold a piece of scrap plywood in front of the trays when scooping potting mix so that no excess falls off the front of the bench. (See the illustration on page 24 to understand how this works.) I considered adding an extra piece of plywood to the front of the bench that would stick up a couple of inches above the seedling trays sitting on the benchtop. This would prevent loose soil from falling off the front of the bench. For now, though, I prefer to have full access to the bench from the front, and I just hold the scrap of plywood when I need it.

Before filling the trays, I slide the lid off the tub and into the corner of the bench, and then I line up three trays in the corner. I hold the scrap of plywood in front of the trays while I scoop potting mix onto them, and then I scrape off the excess soil with the painter's trim guide, leveling the mix with the tops of the trays. I lift the plywood lid slightly, with the trays on top, and drop it a few times to settle the potting mix into the bottom of the cells. Holding the scrap plywood again, I repeat the process of scooping and leveling the potting mix, which ensures there is enough mix in the trays. Finally, I press down on the top of the filled tray with an empty tray, making indentations in the potting mix for the seeds to sit in.

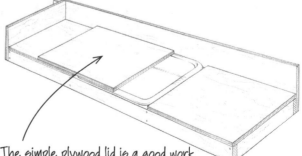

The simple plywood lid is a good work surface, both when covering the mixing tub and when slid over to the corner.

CHAPTER 3
TOOLS FOR THE FIELD

THIS CHAPTER HAS DESIGNS FOR THREE TOOLS that range from extremely simple to the only project in the book that makes use of basic steel fabrication techniques such as cutting steel and welding. As with most of the projects in the book, the designs in this chapter start with tools that are available commercially and modify their forms to fit the day-to-day needs of the farmer, paying particular attention to ergonomics.

HOOP BENDER

On the farm, and even in the garden, it's handy to use hoops to hold up floating row cover, insect netting, bird netting, and other coverings to protect your plants. Different applications require different hoop sizes, but you can easily and inexpensively build a form to bend customized hoops from 9- or 10-gauge wire, electrical metallic tubing (EMT conduit) of most sizes, or even heavier steel pipe such as top rail fencing.

The design shown here is for a hoop bender that will form low, 30-inch-wide hoops from ½-inch EMT conduit. I find this size convenient for protecting early spring seedlings and shorter crops on my 30-inch-wide bed tops. If desired, you can adjust the measurements given here to build a bender that makes a smaller or larger hoop and that works with materials of different diameters.

PROJECT OVERVIEW

Approximate materials cost:
$20

Time to build: 30–60 minutes

Level of complexity:
Moderate

SUGGESTED TOOLS

- Jigsaw
- Drill

RECOMMENDED MATERIALS

Two scraps of ½" plywood, roughly 15" × 15" and 20" × 20"

One scrap of 2×2 lumber, about 6" long

⅞" screws

1½" screws

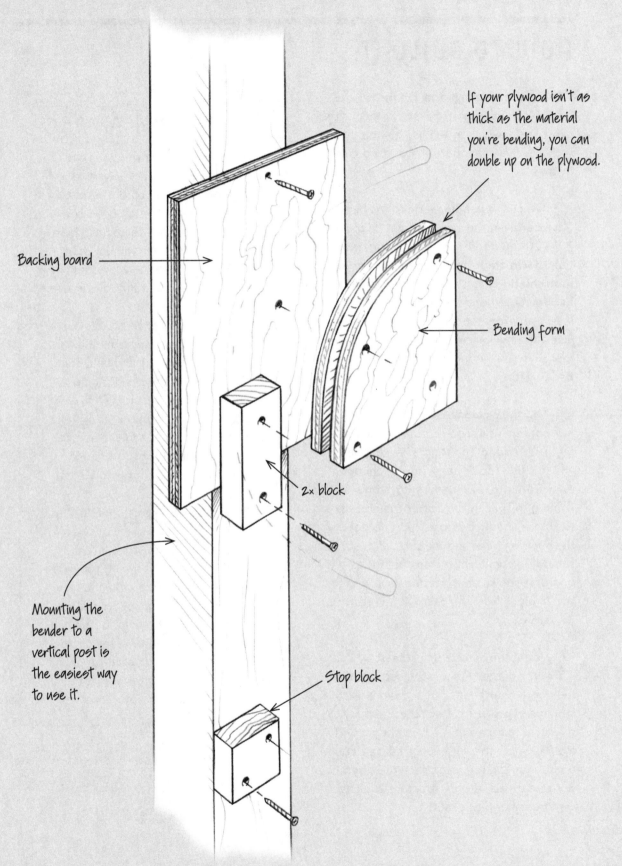

If your plywood isn't as thick as the material you're bending, you can double up on the plywood.

Backing board

Bending form

2x block

Mounting the bender to a vertical post is the easiest way to use it.

Stop block

HOW TO BUILD IT

1 Cut the bending form. On the 15" × 15" piece of plywood, use one corner as the center from which to scribe a quarter circle with a radius of about 14½". Cut the arc with a jigsaw.

2 Attach the bending form to the backing board. Set the bending form on the 20" × 20" plywood, following the orientation shown on the facing page. The bending form should be flush with the edge of the backing board, and its corner should be at least 1" up from the backing board's corner. Screw the two pieces together, using at least four ⅞" screws around the perimeter of the bending form.

3 Add the 2× block. Hold a piece of ½" EMT conduit as a spacer at the end of the arc on the interior of the square backing board, and position the 2× block just far enough away from the bending form to allow the conduit to slide in and out. The center of the block should be close to the corner of the bending form, and the sides of the block should be parallel to the edge of the backing board, as shown on the facing page. Attach the 2× block from the back with two or three 1½" screws.

4 Mount the bender. Use at least two or three of the 1½" screws to secure the bender to a post in a barn or a similarly solid structure. I prefer to attach the assembly to a vertical surface about 48" off the ground in a place where there is enough space to bend the tubing. You can also attach the assembly horizontally to a sturdy structure, such as a workbench or picnic table.

TIP DRAWING AN ARC

An arc is basically a portion of a circle. And there are a few easy ways to draw a perfect circle on a board. One of the most common ways is to drive a small nail into the board, at the point where the center of the circle will be. Then tie a string to the nail, leaving the knot loose so the string can slide around the nail.

Mark the radius of your circle or arc on the board, measuring from the nail. Set your pencil tip on that mark, hold the string tight to the pencil, and use the pencil to scribe an arcing line while maintaining tension on the string; this keeps the pencil the correct distance from the center for the full circumference of the circle, or for the full length of the arc.

When I'm working with wood, I'm more likely to have a piece of scrap wood than a piece of string. In that case, I drive a nail or screw through the scrap wood (1×2 works well, and I frequently have scraps from old stakes). I then hold down the point of the nail where I want the center of my circle and scribe my line by holding a pencil against the edge of the wood at the correct distance from the center.

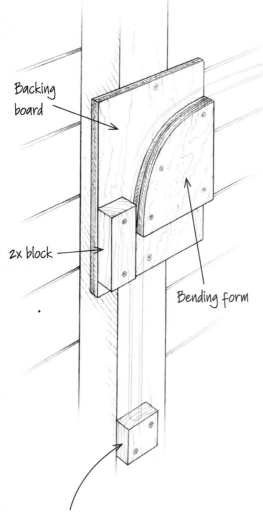

Backing board

2x block

Bending form

A stop block can be placed below the bender to help mark the location for the first bend. For a 5' length of conduit, place the stop block 2½' below the corner of the arc.

BENDING AND MEASURING

When designing a hoop bender, you'll need to consider the properties of the materials you'll be bending and the diameter of the material relative to the radius of the bend. For example, ½-inch EMT conduit is fairly malleable but its wall is thin compared to its diameter, so you can only bend it so tightly before it will kink.

The standard pipe bender used by electricians has a radius of about 4 inches, so that gives you an idea of what the lower limit is for simple bending of EMT conduit. Larger-diameter materials require larger bending radiuses, and smaller-diameter malleable wire can be bent into tight arcs without breaking. Generally, a fully rounded arc is the simplest shape to make, with the most gradual bend to measure and cut. It's also one of the strongest shapes.

To figure out how long to cut the pipe before bending it to your desired measurements, think of the arc as half of a circle's circumference.

circle circumference = π × circle diameter
π = 3.14 (rounded)

In this case, we know we want the hoop to span 30-inch beds, so that's our diameter, and we can use it to find the circumference and then simply divide the circumference in half to determine the outer measurement of a half circle.

3.14 × 30" = 94.2"
94.2" ÷ 2 = 47.1" for the arc of our hoop

Conveniently, EMT conduit comes in 10-foot lengths. You can cut those in half to get two pieces that are each 5 feet—or 60 inches—long. Each 60-inch piece will make a hoop with a 47-inch arc and a little more than 6 inches of straight leg on either side. Perfect!

USING THE BENDER

To bend a low hoop, start by marking the middle of a 60-inch length of ½-inch EMT conduit. Align that mark with the corner of the bending form closest to the 2× block. If your hoop bender is mounted vertically, the conduit will be straight up and down. Use your body weight to pull down slowly on the top edge of the conduit while bending it over the form and sliding it against

the backing board; this will keep the hoop from twisting. If you need extra leverage, slide a piece of pipe slightly larger than the conduit over the end of the EMT.

Once you've bent one side of the conduit (making a 90-degree arc), flip it over and bend the other side to complete the 180-degree arc. This will ensure an even, symmetrical bend.

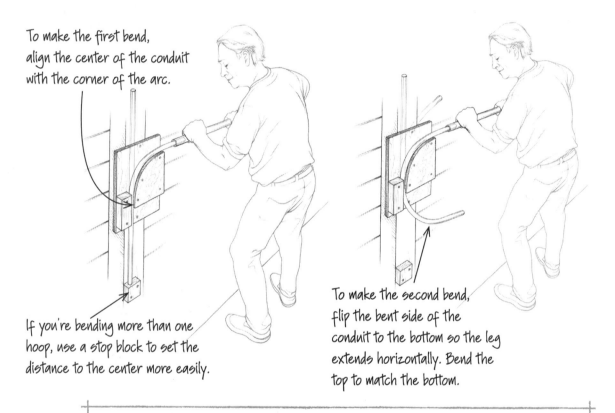

To make the first bend, align the center of the conduit with the corner of the arc.

If you're bending more than one hoop, use a stop block to set the distance to the center more easily.

To make the second bend, flip the bent side of the conduit to the bottom so the leg extends horizontally. Bend the top to match the bottom.

TIPS BENDING THE CONDUIT

- Make all bends in a slow, even motion to prevent kinking.

- The bends will spring back a little bit when you remove them from the bending form. If you need the final hoop to be a little narrower, carefully push the two legs together; if you need the hoop to be wider, pull apart the legs.

- If you don't want to have to measure the center of every piece of conduit, simply measure the first one, put it in place, and screw a block in place to use as a stop when lining up the end of the next piece of conduit.

The Snowmulcher

A Slow Tools Case Study

BRISTOL AGGIE HIGH SCHOOL, DIGHTON, MASSACHUSETTS

Slow Tools is an informal network of North American farmer-engineers created in 2011 by Eliot Coleman and Barry Griffin, with ongoing support from the Stone Barns Center for Food and Agriculture and Johnny's Select Seeds. In 2019, Griffin invited students and teachers from the Bristol County Agricultural High School (Bristol Aggie for short) to attend the eighth Slow Tools annual meeting. Six agricultural mechanics students and two instructors drove five hours in a severe snowstorm to get to the meeting. These were the first high school students ever to attend a Slow Tools event.

After attending the event, the students were full of enthusiasm and ideas, so Griffin handed the Bristol Aggie crew a potential challenge: convert an 8-horsepower gas snowblower to electric power. In addition, he wanted to see if by electrically reversing the fan, the tool would be capable of spreading compost poured into the snow exit funnel. He gave the students a motor and a CAD drawing of a possible fabrication, and the students named this dual-purpose snow/compost machine the Snowmulcher.

Within a month, two students had laid out, hand cut, and MIG welded together a prototype motor frame. With only a few adjustments, the new frame and motor duplicated the footprint, bolt pattern, output shaft height, and position of the original gas motor. They mounted it to a donated snowblower and were ready to tackle the design build. Six weeks and 100 hours later, they completed the design, fabrication, welding, balancing, wiring, testing, and public demonstration of a working prototype. The electric Snowmulcher successfully spread compost!

ROLLING BED MARKER

When planting crops, I appreciate evenly spaced rows and consistent in-row spacing, and not just because it looks neat and tidy. It also helps crops to mature evenly in both size and timing, greatly simplifying harvests and increasing overall yields. In addition, evenly spaced rows make hand-hoeing much faster, as every plant is in a predictable spot and is a predictable size. Unless you're doing all of your seeding and planting with a tractor, marking the beds in advance with both the row locations and in-row spacing allows you to quickly and accurately plant your crops. Using this Rolling Bed Marker is the fastest way I've found to mark out the rows and the in-row spacing.

This bed marker has plywood rounds that mark the rows and four hardwood 1×1 strips that create cross-marks every foot as a guide for in-row spacing. The 1×1s are secured to the wheels with 2×2 braces. A handle-and-axle assembly completes the project.

You can make the handle with wood for a lightweight version of the tool or with either black or galvanized plumbing pipe, which adds some weight and heft. The pipe version has axles similar to those on the Drip Winder on page 76, and it works particularly well in conjunction with the Hand Cart on page 64. Using the two tools together will allow you to mark out pathways and planting rows at the same time. The lighter weight of the wood version is advantageous when you're carrying the bed marker from the tool shed, but it can be a bit of a disadvantage in the field because it doesn't always press down hard enough to make clear marks in the soil.

PROJECT OVERVIEW

Approximate cost: $25
Time to build: Half a day
Level of complexity: Moderate

SUGGESTED TOOLS

- Jigsaw
- Drill
- Framing square

RECOMMENDED MATERIALS

½" or ¾" plywood

2×2 boards

1×1 hardwood strips

1½" screws

FOR WOODEN HANDLE VERSION

Two 1×2s, approximately 40" long

One 2×4, a little wider than the distance between your outside rows

Thick wooden dowel 1½" longer than the 2×4 (a closet rod or broom handle works well)

⅜" all-thread rod about 1½" longer than the dowel

Two ⅜" copper pipe couplings

Four ⅜" washers

Two ⅜" nylock nuts

FOR BLACK PIPE HANDLE VERSION

One piece of prethreaded ½" black pipe a little wider than the distance between your outside rows (for the handle top piece)

Two pieces of prethreaded ½" black pipe, approximately 40" long (for the handle sides)

Two ½" 90-degree black pipe elbows

Two ¾" × ½" black pipe reducing tees (see note)

Two ½" black pipe floor flanges

Two ½" × 6" black pipe nipples

Eight ¼"-20 tee nuts

Eight ¾" × ¼"-20 flat-head machine screws (see note)

NOTE: The correct reducing tees can be hard to find, in which case you can just use ¾" black pipe for all of the handle pipes and fittings, but you'll need ½" nipples and flanges for the axle.

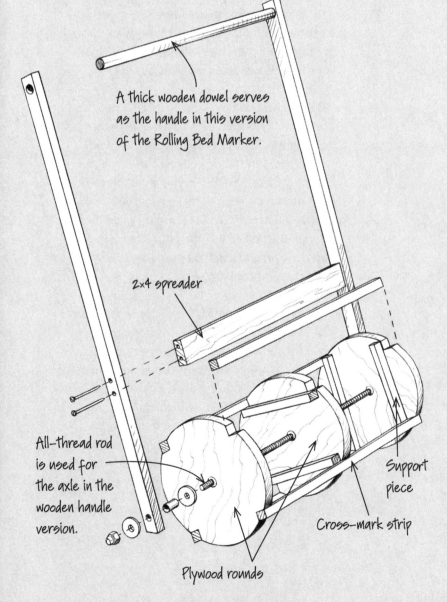

A thick wooden dowel serves as the handle in this version of the Rolling Bed Marker.

2×4 spreader

All-thread rod is used for the axle in the wooden handle version.

Plywood rounds

Support piece

Cross-mark strip

DESIGN NOTES

Determining the size of your marker rounds requires some basic calculations, starting with the desired spacing for the cross-marks. As an example, let's say the cross-marks need to be placed about every 12 inches in the bed. If you were rolling the bed marker on a solid surface, you would make the circumference of the rounds a multiple of 12 inches. And if you wanted to make four cross-marks for each turn of the wheel, the circumference should be 48 inches (12 × 4 = 48). You can then use basic geometry to figure out what the diameter of the circle should be:

Diameter × π = circumference
π = 3.14 (rounded)
Diameter × 3.14 = 48"

To solve for the diameter, divide both sides of the equation by 3.14:

Diameter = 48" ÷ 3.14
Diameter = 15.28"

Now adjust your expectations for real-life conditions. Because you use the bed marker on a soft surface, which lets the rollers dig in a little and slip as they go around, and because the rounds you cut are probably not super precise, you'll find that the final marks are never exactly 12 inches apart. But don't sweat that small error.

The spacing will be pretty consistent, so you'll know approximately how many marks there will be on a full bed, which will be similar to (but slightly different from) the actual length of the bed. I get 70 marks on a 74-foot bed, which means that the cross-marks are spaced about every 12.7 inches. I just treat the marks as if they were 12 inches apart and don't worry about the extra fraction of an inch.

HOW TO BUILD IT

1 **Cut the plywood rounds.** Cut plywood rounds with a radius of $7\frac{5}{8}$". The number of rounds you need depends on how many rows you want your marker to have. (For a three-row marker, cut three rounds.) Your circles don't have to be perfect, but it helps a lot if they're all the same.

To make step 2 easier, drill a $\frac{3}{16}$" hole through the center of each round. If you have a drill press with a deep enough throat, you can stack and clamp all of your rounds together, then drill the hole through all three at the same time. (This will ensure that all of your rounds match as much as possible.)

NOTE: Doing this by hand, without a drill press, is risky because it's easy to angle the drill slightly. That's not such a big deal when you go through the first round, but it will put you far off the mark by the time you hit the third or fourth piece in the stack. If you don't have a drill press, it's most accurate to drill through the first round, then flip that piece upside down and use the hole as a guide to drill each of the next ones separately.

2 **Lay out the rounds.** On each round, draw two perpendicular lines through the center and extending out to the edges, using a framing square. These indicate the locations for the 1×1 cross-mark pieces. Connect the ends of these four lines with straight lines, making a large square. Draw another set of lines parallel to the large square and set inside by half the actual width of your 1×1 pieces. Extend these lines all the way to the edge of the circle.

TIP CUTTING PLYWOOD ROUNDS

An online search will turn up lots of tips and tricks for cutting plywood circles with jigsaws and even with table saws. I've tried a lot of these methods, and if you're going to cut many circles and need them to be near perfect, it might be worth making yourself one of the jigs you can find plans for online. If you do, keep in mind that building the jig accurately and precisely is key to making it work at all.

If you need to cut only a few rounds, it's probably faster to mark the first round using the method described on page 50, and slowly cut out that circle as closely as possible with a jigsaw. Then use the first round you cut to trace the lines for your matching circles before you cut them. (See Design Notes on the facing page for help with sizing your rounds.)

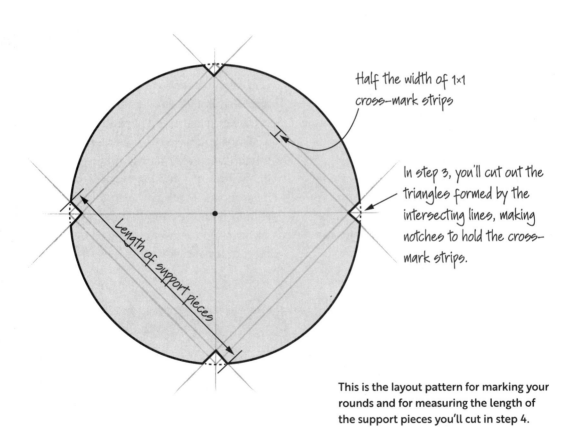

Half the width of 1x1 cross-mark strips

In step 3, you'll cut out the triangles formed by the intersecting lines, making notches to hold the cross-mark strips.

Length of support pieces

This is the layout pattern for marking your rounds and for measuring the length of the support pieces you'll cut in step 4.

3 **Notch the rounds.** Cut out the triangular notch formed by the intersecting lines at each corner of the marked squares. (See the illustration on page 57.)

4 **Cut the support pieces and cross-mark strips.** Cut 2×2 support pieces that are the same length as the second set of lines you made on the rounds (extending from edge to edge on the circle; again, see the illustration on page 57). You need two support pieces per round. Cut the 1×1 cross-mark strips to span the full width of the roller assembly, which should be the distance between your outside planting rows plus the width of your plywood. (My outside rows are 20" apart and I had ½" plywood, so I cut my cross-mark strips 20½" long.)

5 **Drill axle holes.** This step applies only if you're building a version with a wooden handle. Drill a hole through the center of each round, using a drill bit that has the same diameter as the all-thread rod.

6 **Assemble the roller.** This step is a bit fiddly and everything needs to be square so that the rollers won't wobble too much. Take your time and use clamps to hold parts in place as you adjust them before checking their angles and setting them with screws.

First position each pair of support pieces on a round parallel to each other and with their ends flush with the edges of the notches. Screw through the plywood to secure the support pieces in place.

Next, mark the locations of and predrill pilot holes for screws that will go through the cross-mark strips and into the support pieces. To do this, start at the outside rounds and hold the cross-mark strips flush with the outside edges of the rounds while drilling through the cross-mark strips and into the support pieces. The 1×1 cross-mark strips should sit in the notches perpendicular to the round.

Using the pilot holes in the cross-mark strips as guides, screw the cross-mark strips to the support pieces, making sure that everything is square and tight. The roller will be stronger and more stable if the support pieces on each neighboring plywood round are offset by 90 degrees. Both support pieces on the outside rounds need to face inward.

Once you have the outside rounds connected and everything is square, slip the inside round(s) into place based on your desired row spacing. The cross-mark strips should hold the rounds snuggly enough that you can tap them into position before drilling pilot holes and adding screws.

WOODEN HANDLE VERSION

1 **Cut the handle parts.** Cut two 1×2s to length at about 40" long for the sides of the handle. The exact length of the handles isn't critical. In general, taller users will want longer handles and shorter users will want shorter handles.

Cut the 2×4 spreader so it is ¹/₁₆" longer than the width of the roller assembly, plus

the total thickness of two of the axle washers. Cut a piece of dowel the same length as the 2×4 plus the total thickness of the two 1×2 sides (making it approximately 1½" longer than the 2×4).

Cut a length of all-thread rod so it is about 1½" longer than the dowel (see the tip on the facing page). In the case of my roller, where

TIP CUTTING ALL-THREAD

It is easy enough to cut all-thread with a hacksaw or grinder, but inevitably you'll damage the threads at the end slightly, which will make it difficult, if not impossible, to get a nut on from that side. To fix this, simply thread a plain nut onto the good (uncut) end of the all-thread, and thread it off the cut end. The process of threading the nut off the rod will realign the damaged threads on the cut end, effectively repairing it.

Cutting a small slit in the end of each handle side provides just enough room to slide in the dowel.

the outside rows were 20" apart and the roller assembly was 20½" wide, I cut the 2×4 20 11/16" long and the dowel 22¼". I was able to buy a 24" piece of all-thread from my local hardware store that didn't require cutting.

2 Drill the handle sides. Near the bottom end of each 1×2 side, drill a hole just big enough for a copper pipe coupling to slide through. The fit can be tight or slightly loose, but the holes shouldn't be larger than the washers.

At the top of each 1×2 side, drill another hole that is the same size as, or slightly smaller than, the dowel. If the dowel hole is slightly under-sized, you can cut a slit at the end of the handle side piece to allow the wood around the hole to flex open just enough to slide in the dowel.

The location of these holes needs to match on both handle sides.

3 **Install the axle.** Line up the center holes of the roller assembly with the holes at the bottom of the handle sides. Push the all-thread through the center of the roller assembly, capturing a washer between the roller assembly and the side handles to create a little extra space. On each end of the all-thread, slide a ⅜" copper pipe coupling (which will act as a bushing for the all-thread to ride on), then another washer, and finally a nylock nut. Tighten the nuts, but leave a little play in the parts so that everything rotates freely. (The nylock nut allows you to leave a little play in the parts without the nut coming loose over time.)

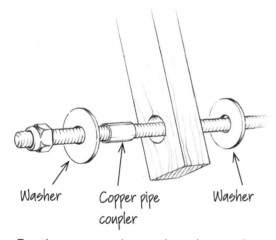

Washer   Copper pipe   Washer
                coupler

Together, a copper pipe coupler and two washers act as a bushing for the all-thread to spin on, reducing wear on the wood.

4 **Fit the dowel.** Insert the dowel into the holes at the top of the handle sides. There may be enough friction to hold the dowel in place without any hardware. If the connection is a bit loose, you can secure the dowel in place with a screw at each end after you add the spreader in step 5.

5 **Add the spreader.** Position the 2×4 spreader between the handle sides, close to the roller assembly but with an inch or two of clearance. Secure the spreader with two screws driven through the outside of each handle side and into each end of the spreader. This will keep the handle from twisting or leaning to the sides.

BLACK PIPE HANDLE VERSION

1 **Assemble the handle.** Thread elbows onto each end of the top piece of the handle. Thread the side pieces of the handle onto the other ends of the elbows, and then thread the reducing tees to the ends of the side pieces. These pieces should fit together snugly, but there's no need to seal the connections with thread-seal tape or pipe dope, or even to use a wrench. The side pieces of handle should line up with each other and make a big U, and the holes through the reducing tees should face each other.

NOTE: The exact length of the handle side pieces isn't critical. In general, taller users will want longer handles and shorter users will want shorter handles.

TIP CUTTING PIPE

You may need to have the pipe for the handle top piece custom-cut (which means you also would have to thread the cut end), but usually you can find a precut length of pipe that will work. Most good hardware stores and plumbing suppliers will cut and thread pipe for you.

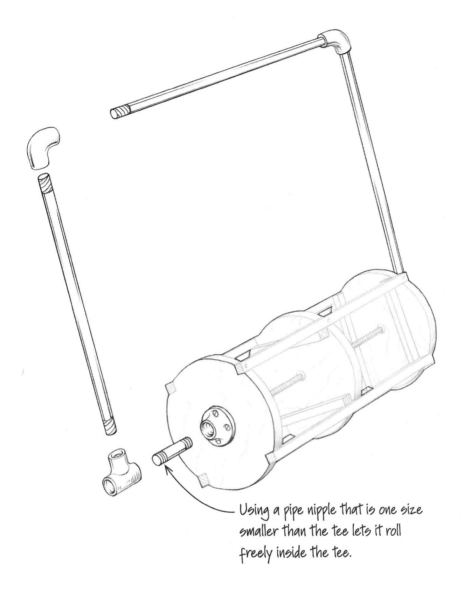

Using a pipe nipple that is one size smaller than the tee lets it roll freely inside the tee.

2 **Mount the flanges.** Center the flanges precisely on the outer rounds of the roller assembly and mark the centers of the four screw holes. Drill holes into the rounds large enough to fit the center stem of a tee nut. Using a hammer, tap the tee nuts into the holes from the inside of the roller assembly, just far enough so that a couple of threads on the machine screws will reach them. Tighten the flanges onto the roller assembly with the machine screws, pulling the tee nuts flush with the inside edges of the plywood rounds and fully sinking the spikes into the wood.

3 **Install the handle.** Line up the pipe tees at the ends of the handle assembly with the flanges, then slide a 6" nipple through the outside of each tee and thread it into the flange on the round. The nipples will roll freely inside the tees. This axle may make noise, but you shouldn't need to lubricate it, and as long as you use the rolling bed marker at a walking pace, the axle should last forever.

USING THE ROLLING BED MARKER

The basic idea behind the Rolling Bed Marker is that you can walk, or drive, down a bed with the tool, and it will mark out the lines on which you need to seed or plant, while placing a mark every 12 inches to help you gauge how far apart to place your seedlings. You could build different rollers that mark different distances, but I suggest using one marker for all of your crops, even if they're being planted at different spacings. Keeping track of and using a separate bed marker for each spacing probably will not save you time.

Using a three-row marker, you can choose to plant one row down the center of the bed; two rows down the outside lines; or all three marked rows. You can also easily plant five rows by planting halfway between each of the three marked rows. With a four-row marker, you can easily plant one, two, three, four, or seven rows (as explained in the table below). These standard row systems simplify marking the beds, planting, placing irrigation, and especially determining the width and setup of cultivation tools.

The Rolling Bed Marker I use now marks four rows on a 30-inch bed top. I can use the row markings to plant between one and seven rows on each bed. If I want just one row, I plant halfway between the two center lines; if I want three rows, I also plant on the outside row lines.

DESIRED NUMBER OF ROWS	PLANTING LOCATION WHEN USING A FOUR-ROW MARKER
1	Between the two middle lines
2	On the outside lines
3	On the outside lines and between the middle lines
4	On every line
5	Plant three lines, plus one line in between each
7	Plant four lines, plus one line in between each

Rolling this marker over freshly tilled soil will mark out three planting rows and make a cross-mark approximately every foot to help guide in-row plant spacing.

In addition to marking out the rows, the Rolling Bed Marker puts a cross-mark approximately every 12 inches. I mark every bed the exact same way, and when it's time to plant, I can apply some simple math to vary the in-row spacing as needed. For example, if I want to plant every 12 inches in each row, I plant at the cross-marks only, but if I want 6-inch in-row spacing, I plant at each cross-mark and halfway between the cross-marks.

In order for this rolling marker to work, you need to start with a fairly flat bed with bare soil. It won't work on a mulch-covered bed, and if the bed is uneven or especially compacted, the tool won't make visible marks. I've found this rolling marker works best immediately after raking, power harrowing, or tilling the surface flat.

I mark every bed we plant, whether we're direct seeding or transplanting, with a few exceptions. I do not mark beds with the rolling marker for leeks and potatoes because we plant both of those crops at the bottoms of furrows. Someday I might make a rolling marker for the bottom of the furrows, but for now I mark where in the bed the furrows will be. I then estimate the in-row spacing, using footsteps to measure as I walk along planting.

If you're using a hand-pulled rolling marker like this in combination with a cultivating tractor, first make the row lines with the tractor. This lines up the rows for cultivation. You can then use the rolling marker to make the cross-marks every 12 inches. Even better, you can use the roller assembly mounted directly to your cultivating tractor to mark out rows; to do so, simply design a mount that replaces the handle.

DESIRED IN-ROW SPACING	PLANTING LOCATION RELATIVE TO CROSS-MARKS
4"	At every mark plus two evenly spaced between the marks
6"	At every mark plus halfway between marks
8"	At the 1st mark and then two marks over, then fill in with two plants evenly spaced between
9"	At the 1st mark, then halfway between the 2nd and 3rd marks, then on the 4th mark, and then halfway between the plants I just planted
12"	At every mark
15"	Once every 1¼ marks
18"	Once every 1½ marks

HAND CART

The inspiration for this design started with the ubiquitous Carts Vermont garden cart, a true workhorse on many small farms and homesteads. I used (and abused) those garden carts for years and have always admired their straightforward design and construction and how functional they are. Despite their excellent durability and hauling capability, I wanted to make some improvements to address specific problems and shortcomings the carts presented around the farm. In particular, I wanted to design a cart that:

- Loads more easily, with less reaching
- Requires less bending over to reach the handle
- Is balanced better, to prevent the load from shifting when I'm moving the cart
- Has more clearance for rolling over planted raised beds
- Uses standard bicycle wheels for smoother rolling

The bed of this Hand Cart is larger and positioned slightly further back than the Carts Vermont bed, resulting in better centering of the load over the axle. The back edge of the bed is about 15 inches behind the center of the axle. This requires you not to put too heavy a load behind the wheel axles (which would flip the cart), but it allows you to transport very heavy loads with relatively little lifting effort. The handle is also longer and stiffer than in a traditional garden cart, giving you a bit more leverage and control when transporting large loads.

The cart wheels are standard 26-inch mountain bike wheels. The wheels do stick up a little over the frame, but I find they don't get in my way. This height puts the bed of the cart at just the right height for easy loading while still providing plenty of clearance over the bed top. I double- and triple-stack bins on this cart and have no trouble moving it, unless the load is more than 300 or 400 pounds or I'm traveling over very bumpy or steep ground.

A narrow strip of wood screwed to the flat bed prevents loads from sliding off the back of the cart.

The optional flat bed sits on the frame and can be removed easily for cleaning, storage, or swapping in other tools.

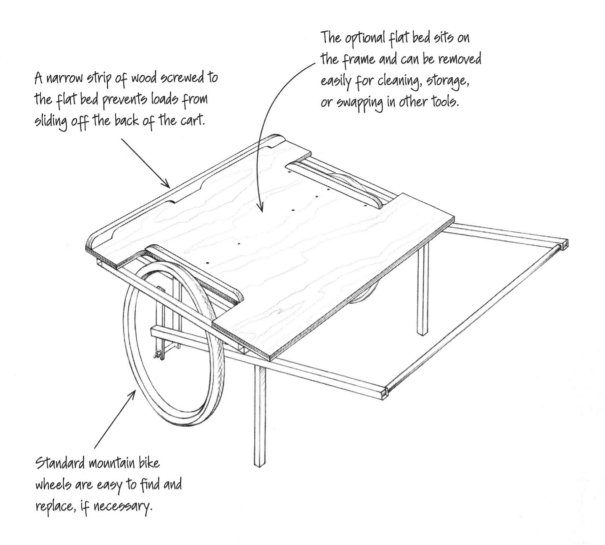

Standard mountain bike wheels are easy to find and replace, if necessary.

NOTE: This project requires basic welding skills and equipment. If you don't know how to weld but would like to learn, this might be a good early project to cut your teeth on (after getting some welding instruction and laying down practice weld beads on off-cuts of metal from the project). Otherwise, you can find someone with welding experience to build it for you—or with you. It's a simple design that uses three basic shapes of plain steel that are easy to weld: flat stock and both square and round tube.

A MIG welder probably works best for this project, but you can also get good results with either a TIG or stick welder. You could also probably braze the whole thing together if you were good with an oxyacetylene torch and had a lot of patience.

PROJECT OVERVIEW

Approximate materials cost: $150–$200

Time to build: 12–20 hours

Level of complexity: Moderate

SUGGESTED TOOLS

- Metal-cutting saw
- Angle grinder or files
- Drill and drill bits for metal
- Sharpie marker
- Straightedge
- Hole saw
- Center punch
- Welder
- Clamps
- Square

RECOMMENDED MATERIALS

Four pieces of 1¼" × 0.250" plain steel flat stock 11½" long (for the fork legs)

Four 1¼" × 0.065" plain steel square tubes, 35½" long (for the side rails)

Two 1¼" × 0.065" plain steel square tubes, the desired width of cart (for the platform front and back)

Two 1¼" × 0.065" plain steel square tubes, 20¼" long (for the stand legs)

Two 1¼" × 0.065" plain steel square tubes, 63" long (for the handle sides)

One piece of 1" × 0.065" plain steel round tube (or up to 1¼" for larger hands, ⅛" shorter than the cart width or ¼" longer than the cart width depending on handle option)

Two 26" mountain bike wheels with tubes and tires

FOR OPTIONAL FLAT BED

⅜" plywood

1×2 lumber

1" screws

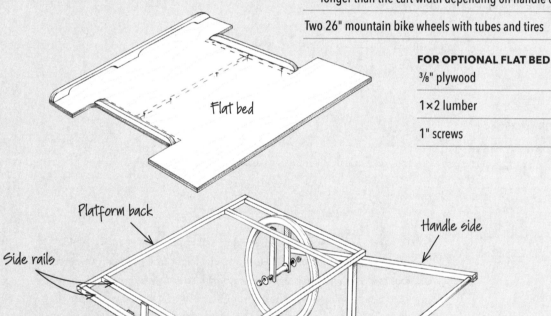

Flat bed

Platform back

Side rails

Fork legs

Slot for the axle

Platform front

Handle side

Handle

Stand leg

HOW TO BUILD IT

1 **Prepare the fork legs.** Use a center punch to mark a hole exactly ½" from one end of each leg, centered on the flat stock. Drill a ⅜" hole at the marked location using a drill press or hand drill. Using a Sharpie and a straightedge draw two lines from the end of the leg to the edges of the hole to outline the slot for inserting the axle. Cut along the lines with an angle grinder or hacksaw. If you cut these slots using a hacksaw, smooth the cut edges with a file. You can also use a file or angle grinder to round the corners of the slot a bit to make it easier for the axle to find the slot.

Cart Frame Dimensions

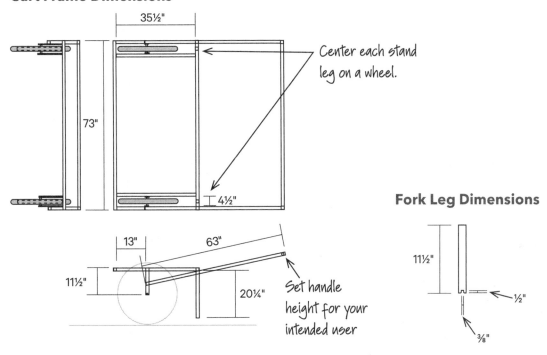

Center each stand leg on a wheel.

Set handle height for your intended user

Fork Leg Dimensions

This cart design is set up for a farm with beds on 66" centers.

NOTE: On design drawings for metal fabrication, it's common to see dimensions with an accuracy of three decimal places when working with inches. Usually this will be accompanied by a plus or minus accuracy number, such as +/- 0.005. This indicates that as long as your measurement is within $5/1000$" on either side (a range of $1/100$" total), the part you're fabricating will work.

For this hand cart, if your measurements are +/- 0.010", or approximately within $1/64$", you'll be fine. In most cases it's more important that parts match each other than that the dimensions are followed exactly. When designing your own parts, think about the level of accuracy needed for the part to work. In other words, how far off the stated dimension can the part be and still work?

2 **Weld the side rails.** On each side rail, mark the location for attaching a fork leg and clamp the two side rail pieces together, using a square to ensure they are at a perfect right angle. Note that the fork legs work in pairs, and in each pair, one fork leg is set 13" in on the inside face of its support, and the other leg is set 13" in on the outside face of its support (see the illustration on page 66). They are mirror images of each other. When you've confirmed the positioning, make two tack welds to hold each pair of pieces together, then check that nothing has shifted. It's important that all four sets are square and exact matches of each other. If one leg sticks out more than its mate, or if one leg is in a slightly different location along the tube, it will misalign the wheel when it sits in the fork.

If the tack-welded pieces are positioned properly, finish the welds by making two weld beads up the sides of the flat stock. If something is a little off when you check the tack welds, sometimes you can bend or tap a piece into alignment. But if it is too far off, grind or file off the tack welds and try again before making the final welds.

3 **Cut the platform front and back.** Cut these pieces to whatever length you want the width of the cart to be. I like to make my carts so that the wheel tracks match the distance between the centers of my beds in the field, which is 66". On the cart shown on pages 66 and 67, the wheel hubs are just under 4" wide and the ¼"-thick fork legs are attached to 1¼" square tube supports. Added altogether for both sides, that makes the full width of the cart 7" wider than the distance between the wheel centers, or 73" total. Drill vent holes in the front and back pieces for moisture and hot air to escape.

4 **Assemble the platform.** To get the platform as flat as possible, clamp the pieces for the outside perimeter of the frame—that is, the platform front, platform back, and two fork supports with inward-facing fork legs—to a flat surface or lay them on a flat slab as you put the assembly together. Double-check that the pieces are square to each other and are sitting flat before tack welding them together. Once all of the outside pieces are tacked together, check that they are still square and flat, and adjust the pieces as necessary before welding the perimeters of the square pipe.

Position the two remaining fork supports inside the frame, with their fork legs facing outward, to complete the forks. Bolt the wheels into place, making sure everything is as square and straight as possible. The wheels should track straight and sit vertically

TIP ADD VENT HOLES

Drill vent holes, or some equivalent, before welding the side rails to the front and back rails. If your welds are continuous, they will totally close off the interior of the square tube sections that connect the front rail to the back rail. As you weld, you are heating the interior air, which creates pressure and is likely to blow out the last bit of weld. Tight welds also mean that any captured moisture from the enclosed air has no way to escape the inside of the rail. This will cause the pieces to fail prematurely from corrosion. To hide the vent holes in a location that bears almost no stress from loads, drill ¼-inch holes on the front and back rails before you attach the side rails.

relative to the frame. Tack weld the two inner supports to the front and back pieces, then check the wheel alignment and adjust as necessary before completing the welds.

5 **Install the stand legs.** Clamp and weld the stand legs to the platform front piece, in line with the wheels. It's nice to get them straight and square, but it's not critical here, so you can do everything by sight. Cap the bottoms of the legs by welding on a piece of ¼" flat stock or a piece of pipe cut to sit just inside the square tube and give a rounded foot. Whichever option you use, drill a vent hole for moisture and hot air to escape.

6 **Prepare the handle sides.** You can cut one end of the square tube side pieces at a 78-degree angle so that it sits flush with the back of the fork (which will give you a handle that is mounted at about the right height for me) or leave the ends square, instead of cutting them at an angle, and adjust the handle height to suit your own needs before you weld them in place in step 8.

7 **Add the handle.** Cut the round tube for the handle ⅛" shorter than the width of the cart if you're using the method that leaves square ends (version A, at right) or ¼" longer than the full width if you prefer a more rounded handle corner (version B, at right).

For method A, use a hole saw to cut a 1" hole in the side pieces that go through only one wall of the square tubing. Cut the holes ⅛" from the ends of the square tubes to fully capture the handle when you insert it into the holes.

For method B, cut a 1" hole through both walls of each side piece, then cut off the ends of the square tubes, lining up your cut with the center of the holes. This makes a

half-round seat for the pipe to sit in. With this approach, the handle connection will rely entirely on the strength of your welds, but it will also create a smoother corner.

Whichever handle version you decide on, slide the handle into the holes (or half-round seats) and tack weld it in place. If you use method B, your handle should extend past each side piece by ⅛" to make it a little easier to weld them together without any blowouts. Make sure everything is flat and square, but complete the first welds in the next step before you finish welding the handle to the side pieces.

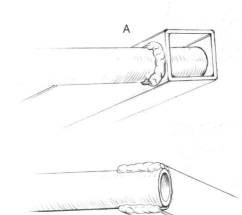

Two options for the handle connections: In version A, the strength of the handle connection is supported fully by the hole in the side handles, and the welds help stiffen the joint. The drawback to this method is that the end of the side piece protrudes slightly. In version B, the strength of the handle connection comes from the welds themselves, and the round tube softens the end of the sides. The drawback to this method is that the round tube sticks out slightly on the sides. Both methods work well.

8 **Install the handle assembly.** Loosely clamp the handle assembly to the underside of the front piece of the platform frame, then adjust the height of the handle to your liking by clamping the end to the outside of the wheel fork. Do this evenly on both sides so that the handle is not twisted. Once the handle is where you want it, make tack welds all around and check that everything is still where you want it, or adjust accordingly.

Make your final welds, beginning by welding across the front edge of the platform frame, then make a bead on top of the square tube and another on the bottom of the square tube where it sits against the wheel fork. Finally, weld together the edges where the round handle contacts the side pieces. The weld connecting the platform frame to the handle assembly doesn't make a super-strong joint for resisting downward force. That's okay because most of the force on that joint is lifting the frame up, which means the joint isn't reliant on the weld at all in its primary mode of loading. That weld will help resist side loads, however, when the handle turns the cart.

9 **Add an optional flat bed.** To build a simple and useful flat bed for the cart, cut a piece of plywood across the 48" width to match, at a minimum, the distance between the two inside fork supports for the wheels (for the cart on page 67, this would be 62", which is the 73" full width minus 11"). This width will ensure that the edges of the bed are fully supported. The frame of the cart is only 38" from front to back, but the plywood is 48", so I overhang it evenly, with 5" off both the front and the back. To keep the plywood centered on the cart frame, cut two pieces of 1×2 the length of the frame interior—just a hair shorter than 35½". Screw these through the plywood deck to the underside so that they sit slightly loose against the sides of the platform.

To keep loads from sliding off the back of the bed, screw a length of 1×2 to the back edge of the plywood. Depending on how wide your cart is, you may want to add a 1×2 piece underneath the plywood to help stiffen it, creating an "H" with the two pieces you've already added for centering it. You can

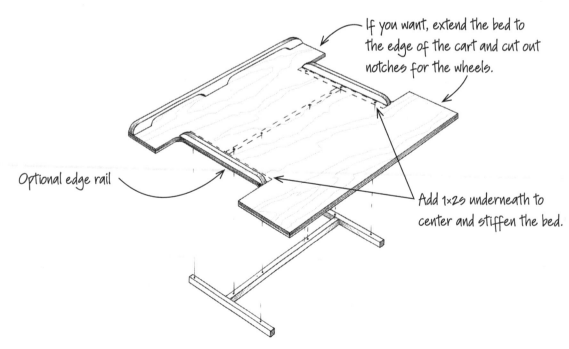

If you want, extend the bed to the edge of the cart and cut out notches for the wheels.

Optional edge rail

Add 1×2s underneath to center and stiffen the bed.

also add strips along the top sides of the plywood to keep harvest totes, or whatever you most commonly carry, from sliding sideways and rubbing against the wheels if you find yourself frequently rolling the cart across side hills.

The flatbed simply sits on the frame of the cart and does not need to be secured to the frame to be fully functional. This makes it easy to remove for cleaning, storage, or replacing with bed styles for different uses or alternative tools.

USING THE HAND CART

I've been using variations on this cart for about 14 years and have built at least a dozen versions for farms around the country. Many other farmers have built their own versions based on my plans and have reported good results. In my personal experience, this cart is a joy to use, and in some ways even more functional than I originally imagined.

At one farm, they built a three-sided box for the cart and filled it with compost. One person would pull the cart and another would take

compost out of the back of the box as the cart moved down the bed. I've dreamed of adapting the drop spreader that we use for soil amendments to take advantage of my cart's larger wheels and greater hauling capacity.

Another variation on the basic design is a single-wheel cart that is a bit like a wheelbarrow, but with a flat bed (shown below). In that version, the load gets centered over a slightly smaller (20-inch) bicycle wheel. The single-wheel cart works better in tight places than the double-wheel cart but can't carry as large of a load. Overall, I find both designs similarly useful.

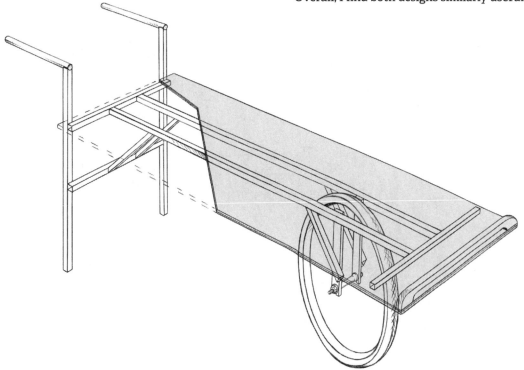

I first built my cart expecting mostly to pull it, as I had with traditional garden carts. In reality, most of the time this cart is a little easier to *push*, and it's nice to be able to see the load as you push. The flat bed also doubles as a nice worktable when you are in the field, making a convenient place to adjust a push seeder and place your seeds, or just to sit and record field notes.

The lack of sides on the flat bed makes it very easy to load bins on and off the cart. Specifically designed to facilitate easy loading and unloading, the height of the deck should be just a little below the bottom of one of our harvest bins when someone is carrying it with straight arms. This keeps the bed high enough that we don't need to bend over to load the cart, but low enough that adding a second layer isn't difficult.

When using the cart, keep in mind that the wheels are the weakest parts of the tool. While they are fairly strong vertically, they will buckle under heavy side loads, such as when the cart has a heavy load and is going around a turn at high speed. If you don't overload the cart, the

wheels will be fine. If you do overload the cart, just make sure the wheels are rolling in a straight line, and avoid going over big bumps or taking turns quickly. If the wheels break, they are easier and less expensive to replace than the cart frame itself.

Over the years I've had ideas for mounting various tools to this frame. The Rolling Bed Marker (see page 54) is one example that I've built and used frequently. For that, I made shorter black pipe handles and added a bracket that drops onto the cart's frame, without requiring tools to secure it (see below). We used the cart with the bed marker for years on a farm that had a tiller on the back of the tractor that would completely flatten out the fields. We'd then push the cart through the flat field, with someone walking directly behind one of the cart wheels to make one pathway while the other wheel would mark out a pathway on the other side of the bed, at the proper spacing. At the same time, the bed marker made planting lines on what would become the bed top.

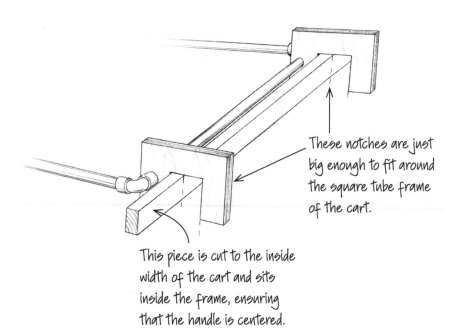

These notches are just big enough to fit around the square tube frame of the cart.

This piece is cut to the inside width of the cart and sits inside the frame, ensuring that the handle is centered.

This simple bracket can be added to the black pipe handle version of the Rolling Bed Marker (see page 54), so that it works with the Hand Cart. The notches in the two plywood end pieces slip over the square tube of the cart frame that is opposite the handle. The long wooden crosspiece matches the inside width of the cart frame, forcing the tool to be centered so that you can line up the cart wheel with the center of the pathway you are pushing it down. This allows you to know that the bed marker is exactly in the center of the bed.

DESIGN NOTES

Often, commercially available tools are designed to take advantage of less expensive materials, and with aesthetics and packaging—rather than the end use—in mind. This introduces into the design a lot of compromises that you can avoid when you design your own tools. In designing this version of the Hand Cart, I wanted to make the proportions fit both the users and the tasks.

FITTING THE USERS. I started by measuring some of the actual proportions of the people working on the farm. I measured the distance from the ground to their fingertips when standing with their arms at their sides, and from the ground to just above the crease at the front of their hips. Then I had them stand normally and hold various harvest crates, and I measured from the ground to the bottoms of the crates.

Setting the handle a couple of inches below an average hip height is a good target height for the design. For large loads, I like to lift the handle to my hips, pushing the load with my hips instead of my arms, and steer with my hands.

Having a flat bed makes the cart much easier to load, since there's no side wall to lift the crates over. The bed is designed to sit pretty flat when the cart is being pulled, resulting in it actually tipping forward a little bit when it is sitting stationary. A slight lip at the back of the bed keeps the load from sliding off. For the bed height, I compromised between creating good ground clearance and allowing users to load the cart without having to bend over too much or lift the crates too high. The sweet spot for both easy loading and good ground clearance was about 20 to 24 inches off the ground.

FITTING THE TASKS. The open frame of this design was inspired by a cultivating tool called the Weed Master. It allows other tools, beside the flat bed, to be mounted on the frame. The flat bed of the cart sits on top of the frame and is held in place and centered by a combination of gravity and simple cleats. I've never needed to bolt the flat bed to the frame, and it's easy to remove the bed for storage.

Because the flat bed is removable, I have made up multiple versions of beds for different uses and can swap them out quickly and easily. For example, I can use one bed for harvest and a different bed, with side walls, for hauling manure or compost.

In addition to measuring the people who would use the cart, I also considered the dimensions of our growing beds and the plants we'd potentially roll the cart over. I decided to mount the wheels at a standard bed spacing so that they roll in the pathways and not on the bed top. I can also remove the flat bed and replace it with the Rolling Bed Marker, as described on the facing page.

Over the years, I have made cart frames with the spacing between the wheel centers as narrow as 36 inches and as wide as 66 inches to match the standard bed spacing on different farms. The design works with all of those frame widths. I simply make the carts to match the width of the farm beds and make the flat beds for the carts to match the width of the cart.

CHAPTER 4
IRRIGATION TOOLS

HERE IN THE PACIFIC NORTHWEST, which is famous for rain, it actually doesn't really rain all summer long, so market farmers rely on irrigation to grow most crops. I have used the three tools in this chapter extensively for more than a decade, the Drip Winder and Drip Irrigation System for more than two decades. These tools are by no means specific to the Pacific Northwest, though. In fact, I got the basic design for the Drip Winder when I worked on a farm in Connecticut many years ago. The instructions for both the Drip Irrigation System and Easy-to-Move Sprinkler System provide a lot of details that will help you design your own system based on the water you have available and the size of the growing spaces in your fields.

DRIP WINDER

Spooling out new drip tape is easy enough to do with only a stick and something to hold it up, but rolling the tape back up for storage, or even just to get it out of the field, is easier with a winder. Larger farms often use power take-off (PTO) winders that quickly retrieve tape onto plastic spools. These winders are expensive, however, and they require a tractor to power them. For most compact farms, a smaller tool that doesn't require a tractor is preferable.

This Drip Winder uses a simple spool made from a piece of plastic pipe, and the winder and stand are made from plywood, a few pieces of 2× lumber, and some steel plumbing fittings. Because the spool is both inexpensive and removable, you can make multiple spools so you have enough to roll up any amount of tape. Plus, when they're not in use, the spools take up very little space. With the winder you can easily roll drip tape onto the spool to make a tidy bale that is compact, easy to store, and just as easy to lay back out in the field next season as a fresh roll would be.

If you like, you can adjust the dimensions of the legs on the stand so that it can sit on the bed of the Hand Cart (page 64) or something similar. That would allow you to easily wheel the Drip Winder around the farm as you roll up your drip tape at the end of the season.

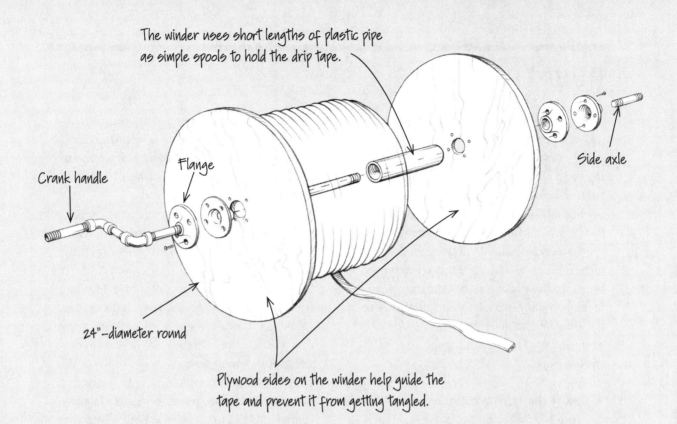

The winder uses short lengths of plastic pipe as simple spools to hold the drip tape.

Crank handle

Flange

Side axle

24"-diameter round

Plywood sides on the winder help guide the tape and prevent it from getting tangled.

PROJECT OVERVIEW

Approximate materials cost: $60–$80

Time to build: Half a day

Level of complexity: Moderate

SUGGESTED TOOLS

- Jigsaw
- Drill
- Hammer
- Framing square or speed square

RECOMMENDED MATERIALS

FOR THE WINDER AND SPOOL:

$\frac{1}{2}$" plywood, at least 24" × 48"

Four $\frac{1}{2}$" black pipe flanges

Four $\frac{1}{2}$" × 6" black pipe nipples

Two $\frac{1}{2}$" 90-degree black pipe elbows

One $\frac{1}{2}$" × 12" black pipe nipple

Eight flat-head machine screws just long enough to fit through two flanges and the thickness of the plywood

Eight tee nuts to match the screws

2"- or 3"-diameter PVC or ABS pipe, at least 12" long

Polypropylene twine

FOR A SIMPLE STAND:

One 2×6, about 36" long

Four 2×4s, each at least 36" long

Two pieces of $\frac{3}{8}$" or $\frac{1}{2}$" plywood, each about 9" × 18"

2$\frac{1}{2}$" deck screws

1" deck screws

NOTE: Sometimes black pipe plumbing parts are harder to find than galvanized plumbing parts. I prefer black pipe because it's a little cheaper and doesn't contain heavy metals. I also like the way it looks and feels, so when I can get it, I do. Galvanized parts work just fine, though, so if galvanized is the only thing available, don't hesitate to substitute it for black pipe. The two are completely interchangeable and compatible, so you can also mix and match if needed.

HOW TO BUILD IT

1 **Prepare the rounds.** Cut two 24"-diameter rounds from the ½" plywood, using a jigsaw (see page 57 for tips on cutting plywood rounds). Drill holes in the center of each round just large enough to let the high portion of the flanges stick through. Using a flange as a template, mark the locations for the four screw holes. Drill these holes just large enough for the tee nuts. Hammer the nuts into the plywood until they are flush. Stack two flanges back to back (flat side to flat side) and screw them into the side of each plywood round, on the face opposite the tee nuts.

2 **Make the crank handle.** Screw a 6" nipple into the flange on the outside of each plywood round; these are the side axles. Make a simple crank on one of the side axles by adding an elbow on the nipple, followed by another 6" nipple, a second elbow, and another 6" nipple. Tighten the connections

by hand or, to get them a little snugger, use a pipe wrench or adjustable pliers. There's no need to lubricate the threads, since none of them will ever need to come apart again.

3 **Make the spool.** Temporarily assemble the winder by screwing both rounds onto the 12" nipple, making them just snug. The insides of the rounds will face each other; the 12" nipple serves as the center axle. Measure the distance between the rounds along the 12" nipple. Loosen the rounds until there are only two or three threads holding them onto the nipple, then measure again. To make the spool, cut the PVC or ABS pipe to a length between these two measurements. Be careful to make this cut precise and square. Mount the spool in the winder by unscrewing one of the rounds from the center axle, slipping the plastic pipe over the axle, then screwing on both rounds so the pipe is sandwiched tightly between them.

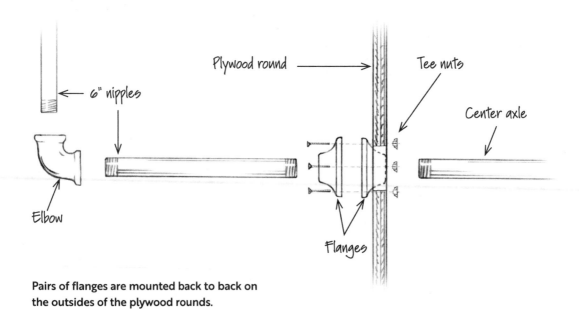

Plywood round

Tee nuts

6" nipples

Center axle

Elbow

Flanges

Pairs of flanges are mounted back to back on the outsides of the plywood rounds.

4 **Make the side axle supports for the stand.** Using a framing square or speed square mark the end of your 2×6 with a 4:12 angle. Flip the square and mark the mirror angle 9" further down the 2×6. Cut out the resulting trapezoid, then flip the cut piece over to use it as a template and mark the next cut on the 2×6 for a matching trapezoid. Depending on the actual width of your 2×6, the long edge will end up being about 12⅔" long. Drill a 1" hole in each piece centered along the shorter side and ¾" in from the edge. Notch these out with a saw so that the side axles can sit down in the grooves.

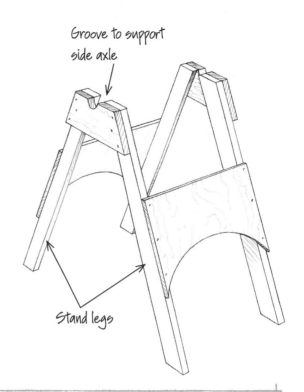

Groove to support side axle

Stand legs

TIP MEASURING A PITCH

When designing tools and marking wood for cutting, it is often more convenient to measure a pitch (rise over run) than to measure an angle.

For example, I knew that the axle on the drip winder stand needed to be about 30 inches above the ground. In addition, I wanted the legs to go from a point that is a bit more than 4 inches from the center of the axle out to a point at the ground a little more than 14 inches from the axle's center line. That's a rise of 30 inches and a run of 10 inches, or 3:1.

To measure that angle, I called for a pitch of 4:12. If you use basic division and divide both sides by 3, you'll see that is the same as 1:3. Turned on its side, a 1:3 triangle is a 3:1 triangle.

To measure a 4:12 pitch, line up one edge of your board with the 4" and 12" marks on a framing square.

This edge is the line you want for the side axle support in step 4.

A 4:12 pitch equals 18.4 degrees.

5 **Make the legs.** Cut four 36" legs from the 2×4s, using the same 4:12 pitch angle on the ends that you used on the 2×6. You can make the measurement on one leg and use that leg as a template to mark the other three for cutting. Screw the legs to the 2×6 side axle supports, using 2½" deck screws and angling outward. Use five screws per leg: one at each corner of the joint and one screw in the center to hold it tightly. I recommend drilling pilot holes for these screws, as they are placed close to the ends and need to be tight and sound, with no splitting. This joint will receive significant torque when the roller is full.

6 **Assemble the stand.** Cut two pieces of plywood long enough to bridge the distance between the two leg assemblies, about 19"–20" and about 9" wide. Screw these pieces to the two assemblies, using at least three 1" deck screws at each joint and leaving an inch or so of clearance for the spool. Position the winder on top of the stand, with the side axles set into the notches of the 2×6 axle supports.

Use the angles you cut on the first leg as a template for marking the cuts on the other legs.

TIP TAPERED LEGS

As an alternative to using 2×4s for the legs, you can rip a 2×6 or 2×8 diagonally to make tapered legs. Tapering puts more material up near the top of the leg, where there are more forces from the joint, and lightens the leg at the bottom, where there is less need for extra support. Tapered legs work well for this stand or for a table.

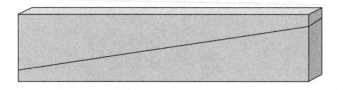

DESIGN NOTES

The original design for this Drip Winder came from a similar tool on a farm in Connecticut where I worked more than 20 years ago. I have no idea who built it originally and I don't have any memory of the stand it was on, but I made something similar for the next farm I worked on, and I've been using one ever since.

For many years, I used the winding mechanism in conjunction with the Hand Cart (page 64), simply cutting and notching two 2×4s so that they sat across the top of the cart and held up the winder. This wasn't a great working height, but having the winder on the cart was convenient in some ways. It made it easy to move the winder from bed to bed as each length of drip tape was wound up, and then to move the whole tool back into the barn when we were done.

You can make a stand like the one pictured at right with legs of any length, allowing it either to sit on a cart or to sit on the ground at a comfortable height for winding without you having to stoop. An ideal stand would position the axle a little below elbow height. A stand with very short legs can sit easily on the flat bed of the Hand Cart.

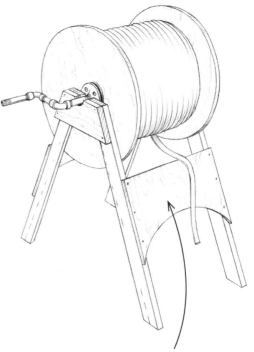

This bridge uses an arc from scraps leftover from cutting the rounds. The arc is optional but does help to reduce stress risers at the joints and makes the whole stand lighter (see page 195 for more on stress risers).

USING THE DRIP WINDER

This is a pretty straightforward tool to use, but there are some things you should know to make it work really well.

New drip tape comes in very compact rolls, and it's impossible to get it that compact again when you wind it back up, but you can come close. To save space on the spools and to keep them from falling apart when they're in storage, wind the tape as tight and flat as possible. To start, insert a new spool in the center of the winder.

Make sure the tape that you're about to wind up is not held down by weeds and that it is open at the far end so that water and air can escape.

Make a full wrap around the spool, then start winding with the crank while you hold a little tension on the tape and pinch it flat to exclude as much air and water as possible. (When you're first learning, it's easiest to do this with two people. One person cranks while the second person guides the tape.) The friction of the tape in tension should keep it from slipping. Work the tape back and forth across the spool as you wind so that the spool fills evenly. If you added couplers to your tape in the field for repairs, remove them as you rewind the tape.

When you come to the end of one piece, pull the last 12 inches or so tight and capture the new tape under it. Slowly start winding again and secure the very end of the new tape under the old tape to keep it from coming loose. This will make unrolling the tape easier.

If you use loops of drip tape (similar to what I describe on page 91), roll up the tape starting at the far end of the bed opposite the manifold. This will roll up the tape from the center of the loop and make the process go quickly, since you're essentially rolling two bed lengths of tape at once.

The spool is full when you are within an inch of the edge of the plywood round. Before removing the full spool, you will need to tie the tape

tightly to the spool in at least three locations to ensure that you can reuse the tape without hours of tangled frustration.

Start by lifting the winder out of the notches and place the winder on its side on top of the stand. Unscrew the top plywood round from the winder, exposing the roll of tape. Drop the free end of a length of polypropylene twine down the inside of the spool and carefully lift the bottom edge of the tape roll just enough to pull the twine through, looping it around and tying it tightly. Repeat this tying process in two or three more places around the spooled tape to make a tight bale. Handle these bales carefully and store them on their sides so that they do not come apart at the edges.

When you need to use the tape again, remount a bale on the winder, cut or untie the twine, and roll the tape back out. For easiest use, if you have different lengths or types of tape, wind each kind on a separate spool and label the spools clearly with something that won't rot or fade. The neater you are with the whole process, the better the drip winder will work.

I also roll up tape before bringing it to the dump or recycling it. To recover the spool centers before you dispose of the tape, saw through a bound-up bale with a sharp knife until you free the center. Then, retie the cut tape pieces into a bundle.

Tie your bales tightly for storage to prevent them from coming apart and becoming tangled.

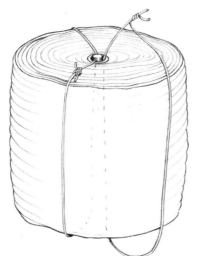

When you're ready to dispose of tape, cut the spool out of the center. This not only saves the spool but also gives you a compact bundle of short pieces to discard.

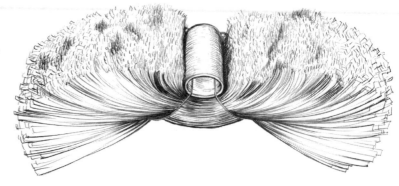

Connecting Farmer to Farmer

The Importance of Finding Good Mentors

I met Rohn Amegatcher at a panel discussion about Black farmers in the Pacific Northwest. During the panel he told a story of the value to farmers of finding mentors "with a crafter's mind" who are willing to spend time puzzling through challenges and helping to pass along the keys to new skills, with the necessary patience.

Rohn farms in Washington State, but when he wasn't able to find anyone locally to help him repair his aging compact tractor, through a network of connections he got in touch with an older Black tobacco farmer in Tennessee who took the time to walk him through repairs over the phone. For that tobacco grower, self-reliance was essential because, as a Black farmer, he found he couldn't trust local mechanics to show up when he needed them; they were just as likely to leave him hanging until the harvest season was over. The farmer in Tennessee was a strong advocate of learning to use hand tools and the craft of hands-on repair, no doubt based on his long practical experience with all the demands of day-to-day field work.

I've found similar thrift and ingenuity among experienced farm workers from Mexico and other parts of the world where there are fewer options for buying a fix than there are in the United States. I've learned a ton working alongside these experienced problem solvers. Often, seeing their solutions to problems and their corresponding tool creations makes me realize how simple and plain a tool can be, and that fancy materials or construction techniques are rarely required.

In today's world, where there are seemingly endless YouTube videos showing all sorts of repairs, construction techniques, and tools—not to mention books like this one—it's easy to lose track of the benefits of personal mentorship and good networks of actual people with practical experience. The farmer I apprenticed under was always willing to help other farmers who asked for help because he understood that we are all interdependent. His example taught me that in-person connections are an invaluable part of learning. I encourage you to make it a practice to share your own ideas and experiences freely and humbly; the favor will invariably be returned.

DRIP IRRIGATION SYSTEM

Water is crucial to plant growth. And ensuring that your crops always have the water they need for optimal growth, whether it rains or not, can make a huge difference in the quality, consistency, and yield. A well-designed irrigation system is a powerful tool for making water evenly available to plants when they need it, with minimal effort and expense.

This project provides an overview of the basic drip irrigation system we use on our farm and walks through the elements of designing a system, calculating water needs, and choosing the right materials and parts. Your drip system may look very different from the one shown here, and therein lies the beauty of drip irrigation: It is highly customizable as well as easy to work with and relatively inexpensive.

Building a good irrigation system is not difficult at all, but it can take a while because there are a lot of little measurements and connections. Even after decades of putting irrigation systems together, I still run into unexpected twists each spring.

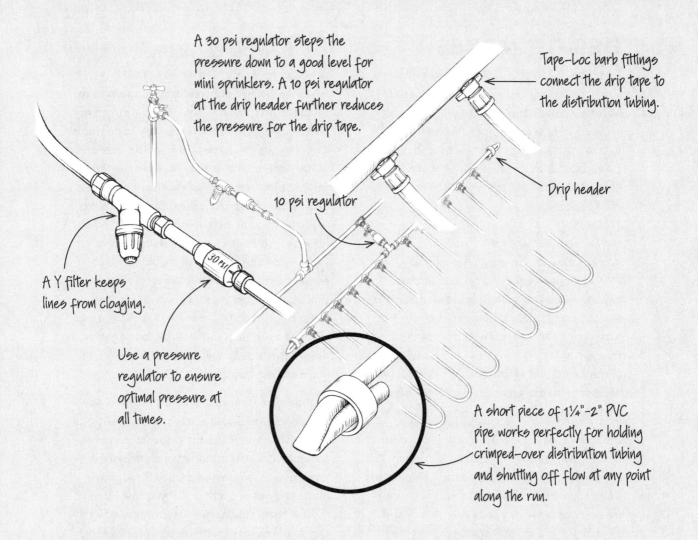

A 30 psi regulator steps the pressure down to a good level for mini sprinklers. A 10 psi regulator at the drip header further reduces the pressure for the drip tape.

Tape-Loc barb fittings connect the drip tape to the distribution tubing.

Drip header

10 psi regulator

A Y filter keeps lines from clogging.

Use a pressure regulator to ensure optimal pressure at all times.

A short piece of 1¼"-2" PVC pipe works perfectly for holding crimped-over distribution tubing and shutting off flow at any point along the run.

PROJECT OVERVIEW

Approximate materials cost: $200+

Time to build: Half a day

Level of complexity: Moderate

SUGGESTED TOOLS

- Bypass pruners or sharp knife
- Drip irrigation punch
- Reel tape measure

RECOMMENDED MATERIALS

Distribution tubing

Drip tape

Tape-Loc fittings to match the tubing and tape

Pressure regulator to match the drip tape

Y filter

6" ground staples

DESIGN NOTES

To design an effective irrigation system for my farm, I had to understand not only how much water my crops need but also the factors contributing to how those crops receive water. To determine how much and how often to water, I started with information from books and then used a lot of my own trial and error and careful observation.

WATERING FREQUENCY

As a baseline, and to greatly oversimplify, I apply 1" of water per week but adjust up or down depending on the weather and the maturity of the crops. I split that 1" of water per week into two separate three-hour waterings because of the type of soil I grow on and the rooting depth of the vegetables I grow. Soil acts as a reservoir, holding on to the water that I put into it (or that it gets from rain).

Because my soil is somewhere in the middle of the sandy to clay range and has decent organic matter, it holds a moderate amount of water (sandy soils drain well and don't hold a lot of water, while clay soils drain slowly and can hold large amounts of water). If my soil had more water-holding capacity, I might go a little longer between waterings, and if it had less water-holding capacity, I might need to water more often. Similarly, plants with shallower roots access less soil, and therefore less water, and may need more frequent watering than deeply rooted crops, which can go longer between waterings because they have access to more soil and more water.

IRRIGATION SYSTEM CAPACITY

When designing your system, it's important to calculate how much water will run through each part of the system. The easiest way to do this is to work your way backward from the field end of the system, calculating how much water it will put out. Then, look at each connection to determine what size tube, fittings, filters, and regulators you need to handle the anticipated flow rates through each part of the system. If you undersize any part of the system, not enough water will be able to get through and some part of your field will end up with less water than it needs. Oversizing the system usually isn't as harmful to the crops, but it can be more expensive, wasting not only money and materials but also water.

In addition to knowing the flow rate and operating pressure of your water source, when you design your system you should also consider potential changes in water pressure caused by the piping layout. As water runs through the pipes and hoses, there is always some internal friction that reduces the water pressure slightly. Larger-diameter pipes have less pressure loss than smaller-diameter pipes because water moves more slowly through a larger pipe for a given flow rate, and friction losses (the engineering term for losses due to friction) increase with speed. Friction losses also add up along the length of the pipe, so shorter runs of pipe have less pressure loss than longer runs.

There are also pressure losses any time water goes through a change in the pipe geometry, such as at a curve, elbow, or valve. These are typically small losses but they can add up. If your water passes through a pressure regulator and the pipe that comes after the regulator is too small or contains too many restrictions before the water gets to the drip lines, the pressure won't be what you expected by the time it gets to the field. For this reason, I wait to put in a pressure regulator that reduces the pressure to a level that works for my drip tape until the water gets to the drip header, the manifold that feeds the drip tape lines. I like to put a valve there, too, so that I don't have to walk back to the source to turn off the water if I need to make repairs in the field while the water is turned on at the source.

Changes in elevation will also change the pressure. If your field is sloped, and the regulator is at the low end of the field, you might not have enough pressure at the top or high end of the field. Whenever possible, I place my regulator at, or near, the top of a field that slopes. There will always be some small natural reduction in pressure at the far end of a field, so even if the elevation is lower, friction losses will slightly offset the increase in pressure due to elevation drop. (For more information on pressure losses due to elevation, see page 196.)

If you're interested in seeing how much friction loss there is in a pipe, there are helpful charts online that relate pipe size, flow rate, distance, and pressure loss. These are particularly helpful if you need to transport water long distances through pipes.

HOW TO BUILD IT

1 **Determine your irrigation needs.** As a first step in designing an irrigation system, it's important to understand what your crops need, including how frequently each crop needs to be watered and how quickly the water should be applied. In addition to informing your watering schedule, this will help you determine the best irrigation method for a given crop (some crops prefer to get their leaves wet, others don't; some need more of the surface soil wetted, others need water delivered to their deeper roots).

As a very rough average, vegetable crops use about 1" of water per week. There are, of course, wide variations in this average. Crops need much less water when they are just germinating and when the weather is cool and damp, but they need up to three times that rough average in a week when they are mature and the weather is very hot, dry, and windy.

The **soil type** will also affect how frequently you should irrigate, since different soils respond to water differently. Sandy soil holds relatively little water, while clay soils hold quite a bit of water. In addition, soil has an **infiltration rate**, which indicates how quickly the soil absorbs water. Sandy soils and high-organic-matter soils tend to have very high infiltration rates, while clay soils—even though they ultimately will absorb more water—have very low infiltration rates and need to be watered more slowly. If your water application rate is faster than the soil's infiltration rate, the water will puddle or run off.

The **rooting depth** matters because each cubic foot of soil will hold a finite amount of water. As you apply water to the soil, it soaks downward from the surface. The water travels down deeper into the soil only when the layer above becomes saturated. If you put on too much water at one time, it will travel below the rooting depth of the plants, which potentially washes valuable nutrients away from the zone where plant roots can reach them.

If you're getting to know a new soil or you want a better read on how much water is available to plant roots in a bed, take a 9"–12" core sample with a soil sampling probe. This lets you see and feel what's happening well below the soil surface. Give the soil a firm squeeze. Soil that is completely dry will fall apart, whereas wet soil sticks together. You never want the soil to dry out completely between irrigations. It should stick together a bit when you squeeze it.

How easily the soil crumbles after squeezing will give you an indication of how much water it contains, and that will vary quite a bit depending on the soil type. In general, if the soil forms a ball when you firmly squeeze it but the ball shatters when tapped firmly, it is ready for its next irrigation.

TIP MATCHING YOUR DESIGN TO YOUR SOIL

It's important to match your water application rate with the characteristics of your soil. A faster application rate tends to spread out more, and in some soils with low infiltration rates (such as heavy clay) this might create pooling on the surface. Lower application rates will water a narrower strip, but the water tends to go deeper for a given soil.

Even though I have a high-infiltration-rate soil and I'd like the water to spread more, I've found that I can get away with low flow, which means I can run twice as many lines in a single watering "set." This reduces the number of times I need to flip valves and reset my irrigation timer every week. In my soil I've found that spacing emitters 8 inches apart usually provides even water down the line and works well for a wide range of crops. I run two lines of drip tape on a 30-inch bed top no matter the number of planting rows. If you're looking for something different, you can also find 4-inch, 12-inch, and 18-inch emitter spacings.

I still watch my plants and soil carefully for any signs of too much or too little water. Wilting plants would be an extreme sign of too little water in many cases, although crops such as cucurbits will naturally wilt in the middle of a hot, sunny day, even if there is more than enough water available. I mostly look for continuous, healthy plant growth, with no discoloration of leaves. Water is the transport method for most plant nutrition, so even if nutrition is available in the soil, crops won't be able to take it up if there isn't sufficient water available. This will show in the color and size of the leaves.

Too much water is just as bad as too little water, as it tends to cause rot or mold and other problems. The symptoms of too much water often look similar to those of not enough water, since excess water can wash nutrients deep into the soil and out of reach of plant roots. In addition, root rot will prevent the roots from taking up enough water for the tops of the plants and can cause slow growth or even wilting.

WATER VOLUME MEASUREMENTS

You're probably familiar with approximately what a gallon of water looks like, but in irrigation, water is often measured in inches, just as rainfall is measured. In a container with straight sides, the depth of the collected water indicates how much rain fell.

For a given space, such as a bed on a small farm, you can convert that inch measurement to gallons.

1 inch = $\frac{1}{12}$ of a foot
1 cubic foot = 7½ gallons

- If you have a bed that is 100 square feet and you put 1 inch of water on it, that is 8⅓ cubic feet of water (100 feet^2 × $\frac{1}{12}$ foot = 8⅓ feet^3).

- To convert that to gallons, multiply 8⅓ feet^3 × 7½ gallons.

- To put 1 inch of water on the 100-square-foot field, you need 62½ gallons of water.

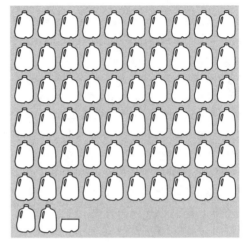

1 inch of water over 100 square feet = 62½ gallons

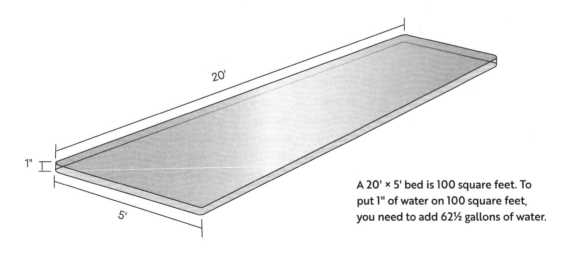

A 20' × 5' bed is 100 square feet. To put 1" of water on 100 square feet, you need to add 62½ gallons of water.

2 Assess your water source(s). Irrigation water can come from many potential sources, including surface water (ponds, rivers, etc.), well water, irrigation canals, municipal water pipes, and rainfall. For most of these, you'll need a pump that moves that water from the source to the field, but in some cases the water may get to the field by simple gravity.

Even if you are using municipal water or another source that doesn't require you to actually pump the water yourself, you'll need to know the **maximum flow rate** and the **operating pressure** for your supply. If you design your irrigation system to never exceed the maximum flow rate, the pressure will remain relatively constant, but once it goes above the maximum, the pressure will drop rapidly. It's important to know the normal operating range and work within that.

Pressure is also affected by the geometry of the pipes, tubing, and fittings it passes through, as well as changes in elevation. If the water is going uphill, through restrictions, or through long lengths of tubing, the pressure will drop. If the water is going downhill, the pressure will increase. Another important factor is the quality of the water, especially for drip irrigation, which relies on many very small pathways in the drip emitters that can easily become clogged if the water is not properly filtered or is high in minerals or algae.

TIP FIGURING OUT YOUR OPERATING PRESSURE AND MAXIMUM FLOW RATE

If you're using a residential water system, it's probably safe to assume that your operating pressure is somewhere between 40 and 80 psi. If you want to know for certain, you can buy a pressure gauge that will screw onto a hose bib and tell you how much pressure you have. Drip systems almost always operate well below residential pressure, so you don't actually need to know the exact pressure. If you run higher-pressure sprinklers, knowing your operating pressure will help you size your sprinklers correctly.

Our farm uses municipal water. The flow rate coming from a standard ¾-inch hose bib is a little more than 5 gpm. We measured this by fully opening a valve and filling a 5-gallon bucket, which took a little under 1 minute.

Note: Flow rate is usually measured in gallons per minute (gpm). Pressure is usually measured in pounds per square inch (psi) in the US or in "bars" where the metric system is used. For flow, sometimes you'll see gallons per minute (gpm) and sometimes you'll see gallons per hour (gph). To convert gph to gpm, which is more useful for design purposes, simply divide by 60 (60 gph = 1 gpm).

3 **Understand the basic flow.** Most drip irrigation systems start with a **filter** at the source of the pumped or gravity-fed water to prevent anything getting through that will clog the system. Next, a system of **pipes or tubes** will carry the filtered water out to the field. (Sometimes the filter is actually in the field, in which case the system carrying the water to the field comes before the filter.)

Once the water is at the field's edge, a device called a **pressure regulator** adjusts the pressure down to a consistent low pressure that matches the needs of the drip lines or tape (where the water is delivered to the soil). A **manifold** coming out of the pressure regulator then distributes the water to all of the drip lines that you want to run. The manifold that feeds the drip tape is commonly called a **drip header**. Finally, you will connect the manifold to the **drip lines**, which can be either straight drip lines that are capped at their ends or looped drip lines that plug into the manifold at both ends.

4 **Choose your materials.** The semi-flexible black tubing made from polyethylene, known as **poly tubing**, is a good option for distributing the water in small-farm irrigation systems because it's relatively inexpensive and durable. It's also easy to work with and reconfigure if necessary. Common sizes are ½" and ¾", with the ½" tubing handling flow rates up to about 4 gpm and the ¾" able to handle double that.

The poly tubing I use is rated for 50 psi. On my farm the water distribution system feeds both sprinklers and drip tape. The sprinklers operate best at 30 psi, so I use a 30 psi pressure regulator to step down the pressure before I transition to the tubing, which keeps the water in the tubing below its 50 psi limit. See the Easy-to-Move Sprinkler System (page 100) for more details on the sprinkler system we use. I add 10 psi regulators at the connection to the drip header manifolds to further reduce the pressure for the drip tape.

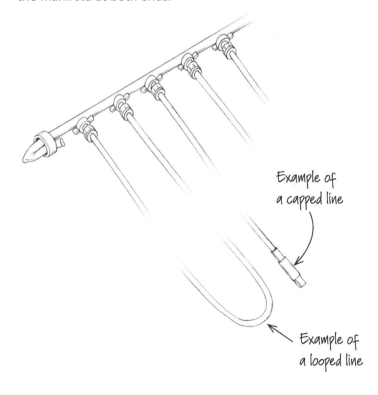

Example of a capped line

Example of a looped line

Running drip lines in a loop means you don't have to cap the ends, and it's also faster to lay them out initially. Looped drip lines have to be run in pairs, and no matter how many rows we plant on a 30" bed top, we always run two lines of drip tape. If you wanted to run an odd number of lines, you would need to cap the end. We did this for years by simply cutting a 1" piece of drip tape, folding over the end of the drip tape twice, and then securing the folds by slipping them into the cut piece of tape.

Irrigation parts are pretty interchangeable, but not everything works together all of the time. A common example is ¾" hoses and pipes. While the threads on ¾" hoses and pipes feel like they might work together, they don't. (See Hose Thread vs. Pipe Thread on the facing page.) If you need to connect one to the other you'll need to get an adapter.

Similarly, poly tubing comes in sizes that seem standard but vary slightly by manufacturer. Tubing and fitting sizes are nominal sizes, which can vary from their actual measured dimensions. For example, not all ½" fittings work with all ½" tubing. I find it easiest to work with a reputable irrigation parts supplier who selects fittings that work with their tubing, instead of mixing and matching from different suppliers and manufacturers.

Fittings for poly tubing come in many styles but fall into two basic categories: compression and barb. With compression fittings, the tubing slides into the fitting, which has a ramped opening that narrows down to an interior ridge just slightly smaller than the tubing diameter. The smaller opening compresses the tube, and the ridge keeps the tube from sliding back out, especially when it's under pressure.

With barb-style fittings, the tube slides over the fitting, which has a ramped ridge (or series of ridges) on the exterior. A separate piece goes over the tubing to keep it compressed over the ridge when it's filled with water under pressure. This separate piece could be a hose clamp or a ring that twists on once the tube is in place. Barb-style fittings are easier than compression fittings to get on and off, but they restrict flow slightly by taking up some of the space inside the tube. Still, I usually use twist-lock barb fittings because they are easy to work with and to reuse.

You may also want to incorporate **valves** into your irrigation system so that you don't have to run the entire system all the time but instead can isolate sections of the farm to water. Because I only have enough water to irrigate one field at a time, there is a valve at each field that I can turn on or off. From year to year I rotate crops in each field, and since I use the drip system on some crops and the sprinkler system on other crops, the type of irrigation I need in each field changes annually, too. The distribution manifold is set up to feed either type of irrigation system and can be reused year after year.

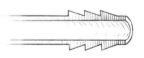

Barb fittings slide into the tubing.

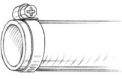

Barb fittings usually require a hose clamp of some sort to hold them on.

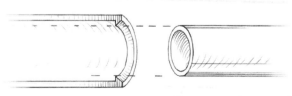

Compression fittings have a single internal ridge and slide over the tubing.

LOOKING CLOSER

HOSE THREAD VS. PIPE THREAD

Hose thread comes in one standard size, ¾ inch. The threads have "parallel" sides and a flat end. They make a seal using a semi-compressible washer (called a hose washer) that sits inside the female threaded end. The male threaded end is flat and presses against the hose washer when it is fully threaded on. When hose connections leak, often it's because the hose washer is either completely missing or it's not seated properly at the end of the threads.

Pipe threads come in a variety of sizes, including ¾ inch, which can almost—but not quite—work with hose thread. Pipe threads don't rely on the male end of the threads being flat and the female end having a shoulder with a washer. In fact, threads on female pipe ends are basically bottomless, meaning there's no shoulder at all. Instead, pipe threads work by tapering, changing their diameter along the threaded section. This results in the threads getting tighter and tighter against each other as you thread them together.

Instead of using a washer to create a good seal, with pipe thread you need to use either thread seal tape (also known as Teflon tape, plumber's tape, or PTFE tape) or pipe dope (a slow-drying, waterproof paste). In general, pipe dope is for metal pipes and thread seal tape is for plastic pipes, although they can both be used on either kind of pipe depending on the specific product and application. Do not use tape or pipe dope on hose threads!

When using thread seal tape on pipe threads, wrap the tape in the same direction that you'll screw the threads together so that it tightens on as the parts are threaded together. If you wrap the tape in the opposite direction, it will remove itself as you screw the parts together.

Regardless of whether you use pipe dope or tape, you'll need to use enough to create a good seal and keep the threads from seizing, but not so much that you clog the threads and prevent them from engaging. Using too much pipe dope or tape can also result in getting some inside the joint, where it will restrict flow or come loose and clog the valves or other fittings.

Pipe threads taper slightly (exaggerated here).

Hose threads have parallel sides.

Drip tape comes in a variety of sizes and configurations. You can also get tubing with inline emitters or use plug-in drip emitters for various applications. Here, I focus on drip tape, which I think is most appropriate for the majority of annual crops. If you're setting up a system for perennials or containers, you might choose a different type of drip, but all of the calculations and considerations will be basically the same.

For a small farm, $5/8"$ drip tape works well. It is the most common size available and comes in different material thicknesses, usually measured in mils (1 mil = 0.001"); different emitter spacings, usually measured in inches between emitters; and different flow rates, usually measured in gallons per minute per 100' of tape.

The material thickness is usually between 6 mils and 15 mils. Lightweight 6-mil material is prone to holes and cuts and often lasts only one season. The 15-mil material is heavyweight and quite durable. It will last multiple seasons unless your water contains a lot of minerals or algae, which may clog it permanently. Because drip tape is sold by weight, for the same price you'll only get roughly one-third the length of 15-mil tape as you would of 6-mil tape. I usually compromise and buy 8-mil tape and expect it to last a few years.

5 **Assemble the pieces.** For best results, start at your water source and work out to your plantings in the field. Essentially you start with the fittings that connect to your water source, then add parts one at a time and measure out tubing as you go, cutting the tubing to length while it sits in its final location. (See A Real Life Example on page 98 for a glimpse at what this process looked like when I put together my system.) As you assemble the pieces of your system keep in mind the following tips.

- **Before screwing together hose threads,** make sure that the hose washers are properly seated at the bottom of the female hose threads and that the connections aren't cross-threaded, which would make them crooked and prevent them from sealing properly. I've occasionally found a female hose thread that is deeper than the mating male threads. When that happens, the male threads don't reach the washer to create a seal. If you find a connection like this, you can stack two hose washers.

- **When making your distribution manifolds,** staple the tees and any elbows to the ground where you want them, using a reel tape to measure if needed. Attach the fitting at the start of your run to one end of your roll of poly tubing, staple that down, and neatly roll out the tubing to the next fitting. (Don't pull the tubing from a stationary roll, as this will add twists and kinks and generally make a mess.) Cut the tubing to length and attach it to the next fitting.

- **To make the drip headers,** which are the manifolds for the actual drip tape, attach the tubing to the tee at the middle of the header (or in the location that fits your design) and staple it down. Roll the tubing out and cut it at least 6"–12" past where you plan to have your last drip tape connection. Staple down the end and lay a reel tape next to the tubing to measure where each Tape-Loc barb fitting should go.

- **Punch holes in your drip headers** where you want the tape to be. As soon as you punch each hole, plug a barb fitting into the hole so you don't lose track of where the hole is. Punches come in different sizes, so make sure you have the correct size for your barb fittings. (See Punching and Fixing Holes on page 96 for more tips on punching holes.)

■ **Be aware that on a long run,** the length of the tube can change by a few inches as it expands in the heat or contracts in the cold. This is also true for the drip tape itself: Usually it's shorter when cold irrigation water is running through it and longer when it's sitting in the sun. To compensate for this and for possible future repairs, add an extra inch or two (or more, for sections that might need to shift in the future) to any long distribution tubing sections.

■ **If you have the option** of warming up the poly tubing in the sun, or maybe in a greenhouse, before getting started, do it. Warm tubing is much easier to work with when making connections and when trying to straighten it out. Having some 6" ground staples to temporarily hold it in place while it takes its new shape also helps. The exception to this, unfortunately, is that it's somewhat easier to punch holes for barb fittings into stiffer, cold tubing. Overall, though, it's easier to work with warm tubing than cold tubing.

■ **Cut tubing** with sharp bypass pruners or a sharp knife. If you use a knife, start by inserting the point into the tubing and then cut around the circumference. Be careful! Irrigation tubing is surprisingly tough, which is good when it is in use, but it can be challenging to cut (and sometimes dangerous, if the cutting tool slips). The ends don't need to be completely square, but it makes things easier if they're close to square and if the cut is clean.

■ **To attach drip tape to Tape-Loc fittings,** squeeze the tape open and carefully push it onto the barb, then screw the locking ring all the way over the tape. This can be tricky to do, but it's worth it to take the time to get the tape on as far as possible. If it doesn't go on the first time, try sticking your finger inside the end of the tape to make it tube-shaped, then maintain that stiff tube as you wiggle the tape over the barbed end. Rotate the locking ring forward so that it covers the tape and the barb without pushing off the tape. Double-check all of the connections once the water is turned on. The tape is actually a little easier to slide on fully when there is water pressure helping to expand the drip tape.

Twist the locking ring to screw it over the drip tape.

Push the tape over the larger barb end as far as possible before rotating the locking ring. To prevent the ring from pushing the tape back, push lightly on the tape as you rotate the ring over it.

The small barb on this end plugs into a hole in the distribution tubing.

Cross section of a Tape-Loc barb fitting for attaching drip tape to the header

TIP PUNCHING AND FIXING HOLES

There are many ways to punch holes in distribution tubing to accept the barb fittings that connect the drip tape and header. I like an adjustable squeeze punch that rotates as it punches the hole and can work on many sizes of poly tubing. I've also used a cordless drill with a small bit and handheld punches that push into the tube.

Using a drill can be tricky, as you're likely to go too far and punch a hole through the opposite side of the tape. Handheld punches are inexpensive and effective but can be slow and tiring if you're punching more than a few holes. Whatever tool you use, it's much easier to punch holes when the tubing is cold and stiff and the punch tip is sharp. You can sharpen the tip with a pocket stone or buy inexpensive replacement tips.

The size and shape of the hole is critical. It's worth spending time lining up the punch exactly where you want it and getting it right the first time. Oversized holes are difficult, and sometimes impossible, to fix without extreme measures. You can plug slightly oversized holes or holes in the wrong spot with a clever, slightly oversized plug called a "goof plug." It's definitely good to have a stash of these on hand. The kind I use has a small barb on one side and a larger barbed plug on the other. The fatter side plugs the hole and the smaller side lets you pull the piece back out if you need to.

If the hole is too big for a goof plug, you'll need to splice in a new piece of tubing with a coupler. It's always good to keep a few couplers and other fittings on hand for repairs throughout the season.

- **To roll out the drip tape two rows at a time,** mount a new roll of tape on the axle of the Drip Winder (page 76). If you don't have a winder, you can use a broom handle held up by sawhorses or a wheelbarrow, or even just set the roll on its side on a piece of cardboard and drive a digging or rock bar into the ground to make a temporary vertical axle. Set up the roll of tape a little behind the drip header, then plug the end of the tape into one of the two Tape-Loc fittings for a bed. Secure the Tape-Loc fitting to the ground with a 6" ground staple to hold it in place while you walk down the bed path, pulling the loop of tape with you. The winder will let the reel unspool as you walk down the bed. At the end of the bed, use two ground staples to temporarily

hold down the tape at the corners until you cut the tape. Return to the start of the bed, cut the tape to length, and plug it into the second Tape-Loc fitting, completing the loop.

- **Flush the distribution tubing** before checking the system for leaks. After all of the tape is laid out, leave the ends of the drip header manifold open and turn on the water. When water starts flowing freely from the ends of the manifold, fold over each end. Secure each fold with a ring of pipe or a few tied-off wraps of twine, then let the system fill to pressure. Depending on the size of your system this can take a few minutes. This will show you if all the fittings are tight and if there are any major leaks.

USING THE DRIP SYSTEM

To run the irrigation system, I set the battery timer on the distribution valve manually. This keeps me from accidentally running the water for too long and wasting it. I also set a timer with an alarm on my phone to remind myself to switch the water to a new field when the valve shuts off.

We have a water meter connected to our irrigation system. It wasn't cheap, but it helps us confirm the quantity of water we put on the field. I like to keep records, so I take a photo of the meter every time I turn an irrigation set on (see Hacking Photos and Videos for Data on page 178). When I'm back in the barn, I note the meter setting and time from the photo, as well as how long I ran the system and what beds I watered. This helps me ensure that I'm watering everything as much as I think I am. If I see problems in a field that suggest a bed is getting either too much or too little water, I can look back and see how much water I actually put on, then adjust as necessary going forward.

I also keep a rain gauge to track the rainfall on the farm between irrigations. If plants need more water because they are mature, the weather is hot and dry, and I want them to grow faster, I sometimes water three times in a week. Or, for deep-rooted crops, I might increase the amount of water I put on in a single watering.

Every time I turn on the system in a field, I pull all of the drip tape lines straight as they fill with water, working from the end opposite the header. Having at least one ground staple over a Tape-Loc fitting on each bed helps hold the header in place while I pull. Then I walk through the field with a long-handled tool such as a hoe, nudging the tape up against the plants I want to water and checking for any holes in the drip line.

If I need to repair a pin hole or small slice, I clean any soil from the area using the water that is escaping and then wrap the still-leaking drip tube twice with vinyl electrical tape just snugly enough to seal the hole. The vinyl tape works well even when wet, and it's easiest to wrap when the line is full of water. I like to use colored tape, not black, so that I can easily see how many repairs are in a line.

If two or three wraps of tape don't fix the hole, wrapping more won't help. Instead, for larger holes or slices, I cut out the damaged section of drip tape and reconnect the two ends with Tape-Loc couplers while the water is on. Working quickly prevents creating a big mud puddle. With a little practice it's easy to get the tape fully onto the fittings when the tube is filled with water.

Even though we're on a municipal water system, I find I still need to clean the filter pretty much every day. I also replace the drip tape every few years when it gets clogged and doesn't water evenly or when it accumulates lots of holes from hoes, harvest knives, and animals. The poly tubing manifolds, headers, and fittings, on the other hand, seem to last forever. We regularly step on them, run over them with tractors, and bump into them with hoes, with no real damage. Occasionally we'll cut one with a sharp, heavy hoe or catch one with a mower and need to make a repair.

I do mow close to our manifold but am careful to not get too close, which means the grass does get high under the manifolds and headers. A few times a year I carefully cut it back with a knife or lift the tubing and mow under it. Some people put strips of weed fabric under their tubing to keep the grass away, but I prefer to not take this extra step because weed fabric can cause some of its own problems.

LOOKING CLOSER

A REAL-LIFE EXAMPLE

SYSTEM LAYOUT. At the field end of our system, each 30-inch-wide bed is irrigated with two lines of drip tape. We've found that two lines of tape set roughly 18 inches apart provide enough water to reach the roots no matter the number of rows planted on the bed. For a single row we run both lines down the bed center. For all other row configurations we run the lines on the outer rows, although when we're germinating seeds or while the plant roots are still very shallow, we sometimes run the first half of the first irrigation set with the two lines closer to the center of the bed to wet more of the surface of the bed. The water spreads out under the surface as it drips, wetting the full width of the bed below the surface.

Our drip tape is listed with a flow rate of 0.34 gpm per 100 feet of tape at 8 psi. Our beds are 74 feet long. With two lines of tape plus a little less than 2 feet making the loop at the far end of the manifold, each bed uses about 150 feet of tape. Given the tape specs, each bed gets about ½ gallon of water per minute, and with 5 gpm available from the hose bib we can run up to 10 beds at a time.

SYSTEM CAPACITY. The filter in our system needs to be able to handle more than 5 gpm. Our drip tape specifies that the filter needs to catch anything larger than 155 mesh (mesh size is the number of openings in a linear inch). We use a basic Y-type filter that is designed to handle up to 13 gpm, which easily accommodates our 5 gpm.

In addition, the pressure regulator needs to work at a flow rate of 5 gpm, and to reduce the pressure to about 8–10 psi to match the specifications for our drip tape. Filters don't usually require a minimum flow rate, but pressure regulators do, so we need to ensure that the minimum flow rate for our pressure regulator isn't more than the least amount of water we'd ever use. In our case, we might want to irrigate just one bed at some point, which would be only ½ gpm, and that will be fine because the working range for our regulator is ½ to 7 gpm.

The hose, tubing, or pipe that feeds the filter and pressure regulator needs to be large enough to handle 5 gpm. As a rough guide, ½-inch tube (approximately equivalent to hose and pipe of the same size) has a large enough internal diameter to handle 240 gph, or 4 gpm. The next size up is ¾ inch, which can handle roughly 480 gph (8 gpm).

Even though ½-inch tubing is less expensive than ¾-inch tubing (especially when it comes to fittings), the smaller tube can't handle the 5 gallons per minute that need to get to the filter and the pressure regulator. Because of that, I use ¾-inch tubing for everything so that I only have to keep spare fittings and tubing for repairs in one size.

We could use the ½-inch tubing for the header, however, if that manifold is split in two. In this case, 5 gallons per minute would need to come out of the regulator and into the header. If the header is fed from the center, with half of the beds on one side of the manifold and the other half of the beds on the other side, it is effectively two manifolds, each serving five beds and each side requiring only 2½ gpm. Feeding headers from the center is good practice for balancing the flow, unless the header runs downhill. In that case, it's usually best to feed the header from the top and let the water run downhill.

SYSTEM SETUP. Our water source comes from a standard ¾-inch hose bib with hose threads. From that I connect a short heavy-duty hose that extends down to a box or platform on the ground where I have a battery-operated irrigation timer. I then thread into a Y-style screen filter, which I can clean out easily, and from the filter into a 30 psi pressure regulator. The timer, filter, and regulator are sized for my maximum flow and all come standard with hose threads, although you can buy them with pipe thread if that's what you need for your system.

Next I use a fitting that has hose thread on one end, to screw into the pressure regulator, and a Tape-Loc barb fitting on the other end, to attach to the poly tube that runs to the field edge. The Tape-Loc fittings incorporate a ring that easily screws over the tubing to hold the tubing tight to the barb.

Along the field edge, I run a long manifold that connects to two valves per block so that I have the option of running half a block at a time. Simple high-flow hose-thread ball valves let me easily switch which block sections I'm running.

When I'm using drip irrigation on a block, I screw a 10 psi regulator to the valve, which tees to a drip header. To close the ends of the manifold, I cut a ring of 1¼-inch plastic pipe, fold over 4 to 6 inches of poly tubing to crimp it off, and slide the ring over the folded tubing to keep it crimped. If I don't have a ring handy, I make a couple of wraps with synthetic baling twine and tie it off.

To connect the drip header to the actual drip tape, I use Tape-Loc barb fittings. These work similarly to the fittings I use when attaching the poly tubing to the drip tape. In that case, a barb presses through a small hole that you punch in the side of the poly tubing. To hold the drip header in place, I add a 6-inch ground staple over the top of every other Tape-Loc fitting. I don't staple down anything else because I want to be able to pull it all up easily for mowing underneath.

Drip tape is designed to be buried just below the soil surface but many farmers leave it sitting on the surface. It's also designed to be installed with the emitters facing up, which helps prevent clogging.

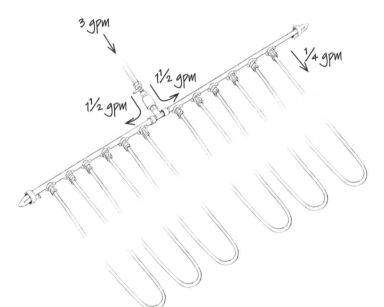

On flat ground, feeding the drip header manifold from the center keeps the pressure and flow to all of the drip lines more even because the distance is reduced from the source to the farthest drip line. Splitting the flow also reduces the speed at which the water is traveling, so friction losses are minimized.

If each drip line is putting out ¼ gpm and there are six lines on each side of the header, the feeder line has to carry a total of 3 gpm. That splits into two flows of 1½ gpm when it reaches the header. Every time the flow passes a line connection in the header, it loses ¼ gpm, until all of the water is accounted for.

EASY-TO-MOVE SPRINKLER SYSTEM

Drip irrigation gets a lot of attention because—when set up properly—it has the potential to use 30 to 40 percent less water than a sprinkler system, and that's important. But drip irrigation isn't appropriate in every situation, and some crops do much better with overhead (sprinkler) irrigation. Regardless of the type of irrigation system I use, I want it to be easy to set up and move (and not cost more than it needs to).

Being able to move the sprinklers easily is important. We irrigate only some of the blocks of beds with sprinklers, and we rotate among different blocks each year to deal with pests, disease, and fertility issues. We also don't plant out full blocks all at once, so where we need the overhead irrigation is always changing. Finally, we need to be able to cultivate and mow for weed control in the same places where we irrigate. With a portable sprinkler system, we can move the sprinkler lines relatively quickly and easily, without lots of extra labor.

Many years ago I saw a K-Line sprinkler system that was designed for irrigating pastures on a friend's farm. Strong, flexible poly tubing connected lines of sprinklers, and each sprinkler stood on a tub-shaped plastic base. The sprinkler lines could be dragged from the end, even while they were running, and moved to a new location. The K-Line system isn't designed for small spaces and it's overly expensive for a compact farm, but I liked the concept. To create my design, I scaled down the basic idea of sprinklers on bases that would slide over the ground and were connected by flexible tubing.

The system I use consists of black polyethylene irrigation tubing (the same material I use for drip headers), weighted sprinkler sled bases, schedule 80 PVC pipe nipples, and pop-up lawn sprinklers, Senninger Wobblers, or Nelson Windfighter agricultural sprinklers. The sprinkler lines are designed to be used with a drip irrigation system, such as the design shown on pages 84–99.

If you're not familiar with drip irrigation, it will be helpful to read through that project to learn the basics of drip irrigation design, materials, and installation and to get yourself started with a new system, as needed. See Design Notes on page 102 for help with sizing your sprinkler system parts and laying out the sprinklers.

PROJECT OVERVIEW

Approximate cost: $25–$30 per sprinkler head
Time to build: 15 minutes per sprinkler head
Level of complexity: Simple

SUGGESTED TOOLS

- Tape measure
- Ground staples
- Bypass pruners or knife

RECOMMENDED MATERIALS

Sprinkler heads (pop-up lawn sprinklers, Senninger Wobblers, or Nelson Windfighter agricultural sprinklers)

Schedule 80 PVC pipe nipples

Weighted sprinkler sled bases

¾" polyethylene irrigation tubing

Male hose thread to ¾" poly tubing adapter fittings

Female hose thread to ¾" poly tubing adapter fittings

Hose ball valve

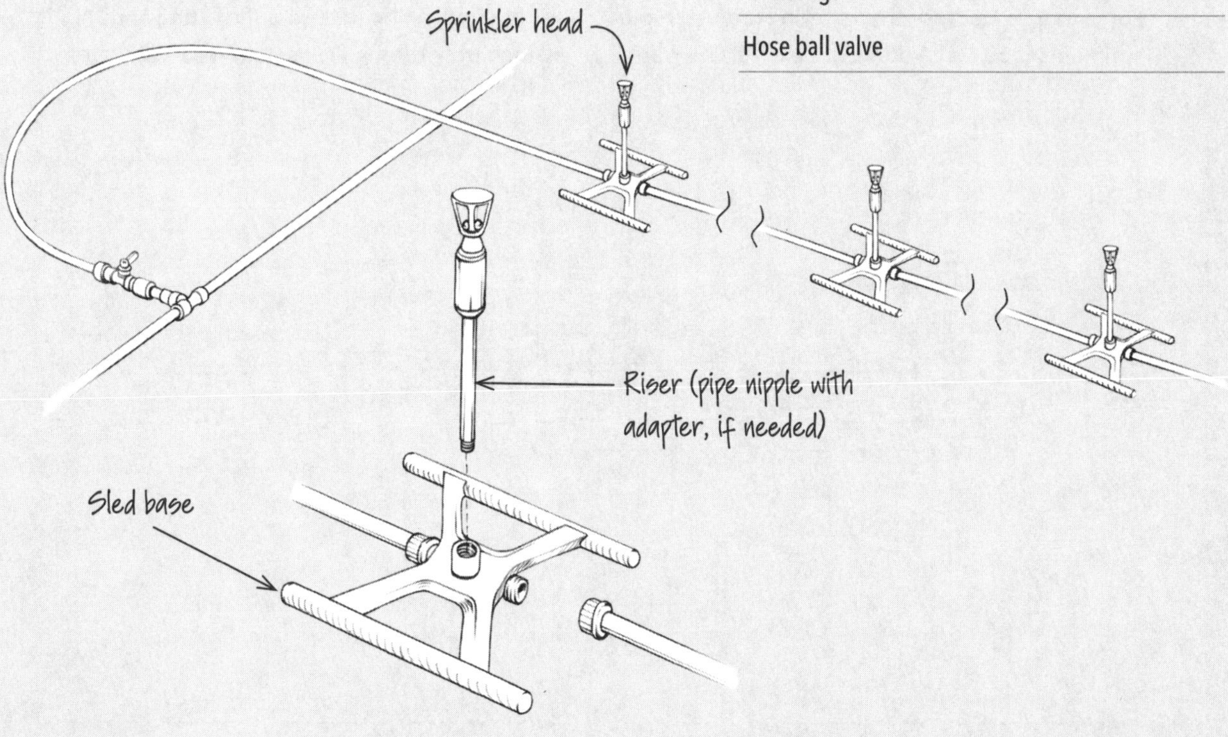

Sprinkler head

Riser (pipe nipple with adapter, if needed)

Sled base

DESIGN NOTES

When designing one of these sprinkler systems, I first consider the sprinkler, since it determines how wide of an area can be watered, and the flow rate of the sprinkler head determines how many heads I can string together. These calculations are similar to those I do for the Drip Irrigation System (page 84). Sprinkler head manufacturers provide charts that list the flow rate for different nozzle sizes at a given pressure as well as the approximate radius that each sprinkler will cover.

At our farm, we use K-Rain ProPlus Rotor sprinklers for some of our overhead irrigation. We set them up with their smallest nozzle and a 30 psi regulator. According to the manufacturer's chart, this should result in a flow rate of about ½ gpm and cover a 28-foot radius. Our municipal water supply offers a total flow of a little more than 5 gpm, which means we could run a maximum of 10 heads at a time. No sprinkler waters evenly out to the full radius that it covers, though.

For example, because the watering radius on our sprinklers is 28 feet, you would expect the water

to reach seven 4-foot beds on either side of the sprinklers. In reality, things do get a little wet 28 feet away from the sprinklers, but I've found that each sprinkler actually only gets enough water to about 3½ beds, or 14 feet, on either side. Any breeze will also cause the windward side to get a bit less water.

To get even water along a sprinkler line, I've found that designing for 100 percent overlap between sprinkler heads works best. This means that, at most, you should set your sprinkler heads about one radius apart. In our case, this means setting the heads 28 feet apart. Just to be safe, we string our heads together about 18 feet apart, which gives us even more uniform water coverage. Because our beds are 74 feet long and the string of sprinklers is about 90 feet, we set the end sprinklers about 8 feet from the ends of the beds. This ensures that the ends of the beds get enough water even if there's a breeze.

In total our six heads use only 3 gpm, and the ¾-inch poly tubing we use allows a flow rate

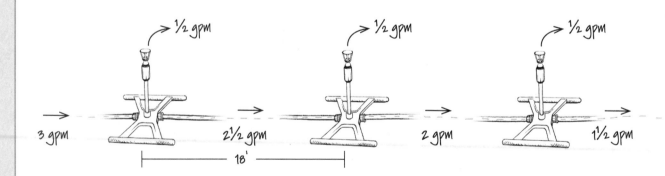

of about 6 gpm. So on one line with six heads, we have plenty of excess capacity. In fact, the only part of the line that even sees 3 gpm going through it is the hose that feeds the entire line. At each sprinkler head ½ gpm escapes, so the tubing between the first two heads sees only 2½ gpm, while 2 gpm goes through the second section, 1½ gpm goes through the third, 1 gpm goes through the fourth, and ½ gpm goes through the fifth.

One reason I use the K-Rain heads is because they have an adjustable rotation pattern, which lets me set them to water 180 degrees instead of 360 degrees. On each end of the bed we have one head set to 180 degrees, and heads in the middle of the bed are set to 360 degrees.

The sprinklers are supported by plastic weighted sled bases that have hose threads going in one side and out the other and ½-inch pipe threads coming out of the top. I use Easy Loc hose thread fittings on the ¾-inch tubing that screws into the sled bases. On the top of the base I attach a 12-inch schedule 80 PVC

nipple (threaded on both ends) as a riser and screw the sprinkler head onto that. The sprinkler heads pop up another 12 inches, which positions them about 24 inches off the ground, well above the height of most crops we use them to irrigate.

Figuring out the best height for the sprinkler heads can be tricky. You need the heads to be high enough to clear the tops of the crops they are irrigating; otherwise, the crops next to them will act as a screen, preventing water from reaching the crops a few beds over. Higher placement also helps the water travel the full radius before hitting the ground.

But placing the heads closer to the ground has potential benefits. It reduces the amount of time the water is in the air, which results in less evaporation and reduced impact from any wind. Lower heads are also more stable and less likely to tip over. The stability of the head is one of the reasons I prefer silent rotating sprinkler heads over "impact" heads, which tend to knock themselves over.

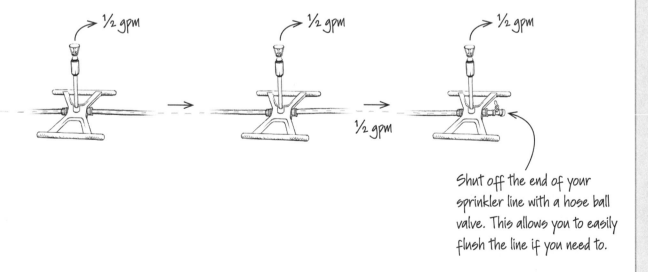

Shut off the end of your sprinkler line with a hose ball valve. This allows you to easily flush the line if you need to.

HOW TO BUILD IT

1 **Connect the sprinkler heads to the bases.** Screw each sprinkler head onto a PVC nipple, then screw the nipple onto a sled base. You should be able to screw both on tightly enough by hand, without any tools. Do not use any pipe dope or Teflon tape, which might clog the sprinkler nozzles later.

2 **Cut the tubing.** Secure one end of the tubing roll, along with the end of a tape measure, to the ground with a ground staple (or you can have a helper hold them). Roll out the tubing, along with the tape measure, to the desired length (see Design Notes on page 102). Cut the tubing with bypass pruners or a sharp knife.

3 **Add the hose fittings.** Fit each length of tubing with a male hose thread fitting on one end and a female hose thread fitting on the other end.

NOTE: We use Easy Loc fittings, which have integrated clamping devices and go on without tools.

4 **Link the sprinklers.** Screw the tubing to the sled bases, linking them all together in a line. Check that each connection has a hose washer before putting it together.

5 **Add a hose ball valve.** Install a hose ball valve onto the end of the final sled base. Having a valve here allows you to easily flush the system to get rid of dirt or insects inside the tubing that might clog the sprinkler nozzles.

USING THE SPRINKLER SYSTEM

I've used multiple versions of this system for more than a decade, and it works great—for a relatively low price. When I need to move the lines, sometimes it's easiest to drag them from one place in the field to another. But with relatively short sprinkler lines, it's easier to roll up the line in 3- to 4-foot-diameter loops, with the sled bases still attached, and then carry the full line to its new location and unroll it in place. When I need to cultivate underneath the irrigation line, I walk down the line, picking up each head and moving the line over by one bed as I go.

The weakest part of the system is the sled base. Originally I used solid, one-piece plastic bases that cost about $5 each. My supplier then switched to bases that are made up of multiple pieces, which over time tend to work themselves loose and cause leaks. For now, I just check the fittings every so often to make sure they're tight and not leaking.

I've seen metal bases that look much more solid but are also much more expensive. At times I have built bases from PVC fittings, but all of the parts end up costing much more than a pre-made base, plus they need to be glued together, and I found they ultimately don't work better than the sled bases.

All things considered, there are a few unavoidable inefficiencies of sprinkler irrigation. One is that sprinklers will never consistently get

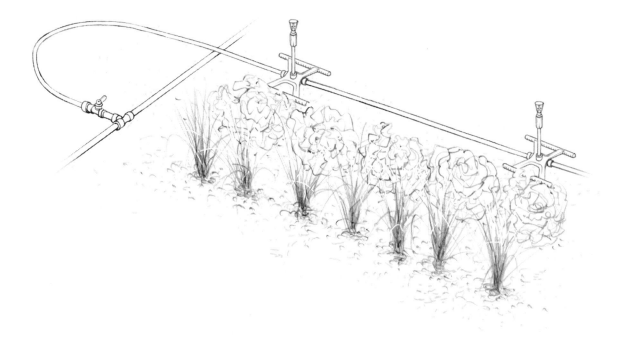

water evenly to every inch of the radius they are watering. Over the years I've noticed that some types of sprinklers apply water more evenly than others, and now I consider that when selecting and running heads. Agricultural sprinklers tend to be more even than consumer and lawn sprinklers, and I've found both the Nelson Windfighters and Senninger Wobblers to be excellent in watering evenly.

I currently use K-Rain pop-up lawn sprinklers, which don't distribute water quite as evenly but do have two other features I like. One is that they can be set to water partial circles, which makes the sprinklers at the ends of the lines more efficient. The other is that they pop up when they turn on, making them taller when we need them to be and shorter when we're storing them.

SMALL WORLD

A few years after I published a version of this design in *Growing for Market* magazine, I was in a local irrigation supply store looking for parts for one of these systems. I started to describe what I needed, and the salesperson said, "Yeah, some guy wrote an article about that and we've had a bunch of folks ask for those same parts." It was very gratifying, but unfortunately they didn't have all the parts.

CHAPTER 5

WASH AND PACK TOOLS

ON THE FARM WE SPEND A LOT OF TIME in our wash and pack area cleaning the harvested vegetables, then sorting and packing them for sale. It's possible to do a lot of these tasks with almost no tools or infrastructure. (When I started my farm, I spent the first month washing and packing on the ground, in the shade of some apple trees.) But it doesn't take much to build some really effective tools that not only speed up your washing and packing process and increase the quality of your produce but also increase your quality of life by improving your working conditions.

Here are a few of the tools I've built and used over the years that I think are worth considering when you're setting up your farm or trying to improve your existing systems.

GLOVE DRYING RACK

In the wash and pack station of the farm, especially during the colder months, we use long gauntlet gloves to keep our hands drier and warmer while we wash produce. Inevitably, at the end of the day the gloves are wet both inside and out. To let the gloves dry passively, they need to be held open somehow to allow for airflow. And since water-saturated air is heavy, it's best if the gloves hang with the openings at the bottom so the heavier moisture-saturated air will drain out.

This easy-to-build rack is made with a single piece of wire that you bend to create "fingers" that hold the gloves. The wire ends slip into holes drilled into wall studs or other supports. The rack sits at an angle to keep the gloves open as they dry.

PROJECT OVERVIEW

Approximate materials cost: $0-$5

Time to build: 15 minutes

Level of complexity: Easy

SUGGESTED TOOLS

- Two pairs of pliers, one with wire cutters (fencing pliers work great)
- Drill

RECOMMENDED MATERIALS

About 96" of 10-gauge soft-core wire to hold six gloves

NOTE: You might also need a block of wood (and some screws) to mount the wire on if you don't have exposed studs or a convenient post or beam for supporting the rack.

HOW TO BUILD IT

1 Shape the "fingers." Hold the wire with one pair of pliers about 9" from one end. Use the second pair of pliers to make a 90-degree bend in the wire. Make two more 90-degree bends to shape a "finger" that is about 6" tall and 1½" wide. Repeat the same process five more times to make a total of six fingers about 1½" apart.

2 Trim the end. After shaping the last finger, trim the wire so that both ends have a straight portion about 9" long.

3 Drill mounting holes. Using a bit that's just a hair larger than your wire, drill a hole into two adjacent wall studs 16" apart. Make the holes at the same height and drill downward at a 45-degree angle about 2" into the wood. Slip the ends of the wires into the holes to secure the rack.

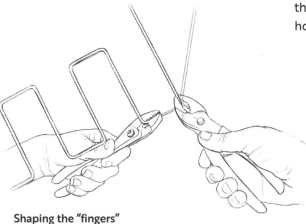

Shaping the "fingers"

USING THE RACK

Simply slide one glove onto each wire finger, allowing the glove to hang with the wrist open at the bottom. This drying rack is a huge improvement over simply tossing your gloves on a table or shelf, it works better than a laundry line (where the gloves tend to hang flat, with no air circulation), and it is well worth the 15 minutes or so it takes to make. The shortcoming of this rack is just that if your barn is cold and damp, the gloves still won't dry particularly well, although they'll be drier than they would be without the rack.

For longer gauntlet gloves, just use more wire and make the wire fingers on the rack longer. If you don't have exposed wood on your wall to mount the rack to, drill holes into a scrap of 2× wood, then mount that wood to the wall.

HAND TRUCK PALLET

Regular shipping pallets work well to move large loads of produce with a pallet jack or forklift, but on compact farms most of us don't work on that scale. This Hand Truck Pallet is a smaller, simpler version that makes it easy to move stacks of produce using a hand truck. It also keeps the bottom of the stack from directly touching the ground, which helps reduce food safety concerns, and improves air circulation below the stack, which is critical to evenly cool your produce.

The pallet consists of a thin, solid platform that roughly matches the 21 × 15-inch base of our harvest containers. Attached to the bottom of the platform are two cleats that run the length of the outside edges. Store the pallets in a dry location when they are not in use.

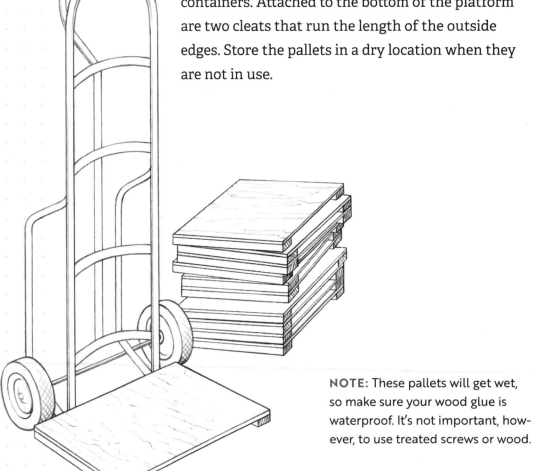

NOTE: These pallets will get wet, so make sure your wood glue is waterproof. It's not important, however, to use treated screws or wood.

PROJECT OVERVIEW

Approximate materials cost: $5 per pallet

Time to build: 15 minutes per pallet

Level of complexity: Easy

SUGGESTED TOOLS

- Table saw or circular saw
- Pneumatic stapler (optional)

RECOMMENDED MATERIALS (for 20 pallets)

One 4' × 8' sheet $3/8$" plywood (CDX works well)

Five 96" lengths of 1×2 furring strips

80-, 100-, or 120-grit sandpaper

Wood glue

$1/4$" × $3/4$" crown staples or eighty $3/4$" screws

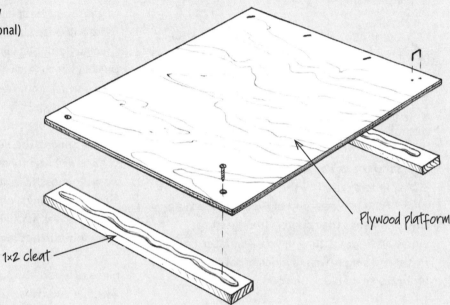

Plywood platform

1×2 cleat

DESIGN NOTES

You can make these pallets from scraps around the farm just as well as you can from new material, and it's not important that they all match, although it is a little easier to stack them if the dimensions are close. It's also not important that the wood for the platform be very stiff, since the cleats on the outside edges of the pallets bear most of the weight when a bin of produce is placed on the hand truck. As long as the plywood is strong enough so that it doesn't warp or sag under its own weight—and will stand up to water, dirt, and the abrasion of sliding harvest bins on and off—it will work. I usually use untreated $3/8$-inch or $1/2$-inch CDX plywood, which is relatively inexpensive, easy to cut, and durable.

For the cleats, you can use any small strips of wood that are at least as thick as the nose of your hand truck and narrow enough to allow the nose to fit under the pallet. The nose on most standard hand trucks is 14 × 7½ inches, but there are smaller and larger versions out there, so check the dimensions of yours before cutting your wood. Because the bottoms of my harvest bins are 21 × 15 inches, I make my pallets just a bit smaller; 19 × 12-inch pallets allow me to get 20 of them out of a single sheet of plywood.

HOW TO BUILD IT

1 Cut the plywood pieces. To make a batch of twenty 12" × 19" pallets from a single sheet of plywood, cut five 19"-wide strips across the 48" length of plywood. Cut each of those strips into four 12" lengths. The exact size isn't critical. For a note on sizing read Using the Pallets on page 113.

2 Cut the 1×2 cleats. For each pallet, cut two 12" cleats from the 1×2 furring strips. You can make these cuts quickly using a miter saw or a table saw with a stop (see Using a Stop Block on page 152), but if you don't have either of those tools, use whatever saw you have.

3 Sand the cut edges. Quickly sand off any splinters raised by the saw blade on the ends of the 1× pieces and plywood. Smoothing the ends keeps the splinters from getting worse and from getting in your way later. For this step, I simply wrap a piece of sandpaper around a scrap block of wood.

4 Attach the cleats. Run a bead of glue down the wide edge of a furring strip and set it flush with one 12" edge of a plywood piece. Secure it with four evenly spaced staples or two screws driven in from the plywood side about 1" from the ends. It isn't necessary to predrill holes in the plywood but doing so will help get the screw heads flush, which is important so that they don't catch on the bins the pallets are supporting. Let the glue dry for 24 hours before using the pallets.

TIP GETTING SCREW HEADS FLUSH

Flat-head screws are designed to sit flush with the surface, and the cleanest way to get this to happen is to start the holes using a countersink bit. You can also get specialty drill bits with a built-in countersink, which lets you drill your through hole and your countersink in one step instead of two.

Alternatively, you can overdrive the screws, sinking the heads below the surface with the force of the driver. This works in most situations and is much faster than using a countersink, but it usually leaves a rough, splintered surface at the hole. In addition, if you overdrive the screw too close to the edge of a piece of wood, the screw head can act as a wedge and cause the wood to split. By overdriving, you're essentially crushing and splintering some of the wood, so it can also reduce the strength of the joint.

In most cases the slight loss of strength isn't important, and splinters will easily sand or chip out. But if clean appearance, consistent countersink depth, and maximum joint strength are important to you, use a countersink bit before driving a flat-head screw.

TIP CUTTING PLYWOOD

A table saw with stands will support the plywood as you feed it into and guide it out of the saw, helping you make long straight cuts accurately. If you have a helper, have them catch the plywood as it comes off the saw.

Alternatively, if you're working with a circular saw, support the full length of the plywood underneath with long strips, such as scrap 2×4s, to set it off the surface and keep it from sagging as it is being cut. Clamp or screw a straightedge to the plywood and make your cuts with a handheld circular saw, running the foot of the saw along the straightedge to ensure a straight cut.

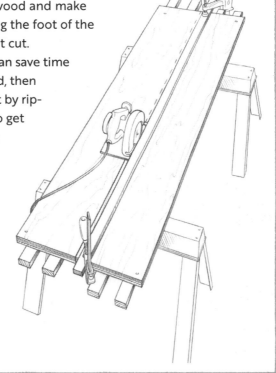

If you need to cut several pallet tops, you can save time measuring and cutting by stacking the plywood, then clamping or screwing the pieces together. Start by ripping the sheet of plywood in half lengthwise to get two 24 × 96-inch strips. Instead of then cutting each strip separately to end up with four 12 × 96-inch pieces, stack the two pieces, clamp them together tightly, set your saw blade a little deeper than normal, and cut through both pieces at once.

Caution: Check that your clamps are secured tightly and that the wood is well supported. Shifting or binding while cutting can cause dangerous kickback.

USING THE PALLETS

These pallets work best on a hard, smooth surface such as a concrete slab. If they're used in soft soil, the cleats tend to sink into the ground, and if they sink unevenly the stacked bins will likely tip over. If you plan to use the pallets on an uneven surface such as wood chips or crushed rock, consider making both the pallets and the cleats a little wider than shown here in order to distribute the load over a larger area.

You can also make the pallets taller to keep your produce bins higher off the ground. And while the added height does offer some benefit, the taller you make them, the more you'll have to tip the hand truck to keep the bottom of the cleat from dragging while transporting a load.

ONION BAG FILLING STAND

I designed this stand to facilitate filling 50-pound mesh bags of onions and potatoes. We pour in 25 pounds at a time from plastic harvest containers, either before or after we clean the crops. We store the onions and potatoes in those bags, stacked on pallets in our walk-in cooler, and either sell full bags or re-portion the contents later in the year.

Only a few dimensions on this design are critical, so it's a great project for fitting together odd scraps of plywood into something useful. What's important is that the top of the funnel matches the width at the top of the harvest tote you're pouring from and that the bottom opening of the funnel matches the mouth of the mesh bag. Both the funnel and the bag sit on a ramp, which slows the onions and potatoes as they tumble downward and prevents bruising. If you want to make a filling stand for a different size of mesh bag, simply adjust the dimensions to match your bag's opening.

DESIGN NOTES

Most of the dimensions in this design are not critical, so you can use up odd scraps of wood instead of going out to buy new material. If you're short on plywood, the base design is particularly easy to change without affecting the functionality. The length and shape of the side wings and the angle of the front of the funnel are also very flexible. I've made several versions of this stand, each with a different base and a different funnel shape (while keeping the openings the same dimensions), and functionally I haven't noticed a difference between them.

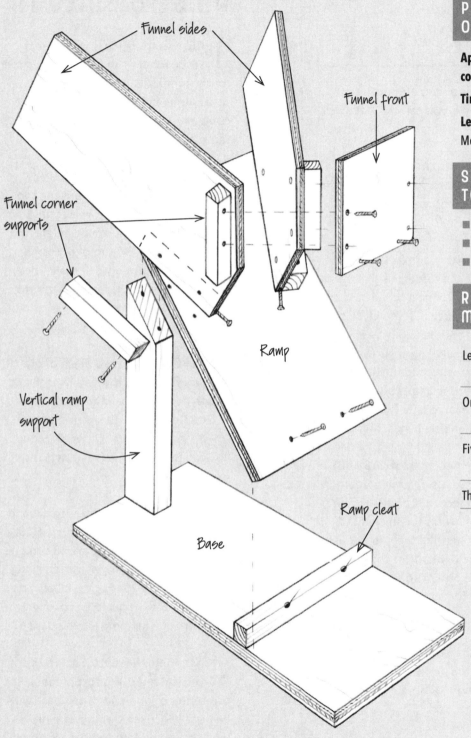

Funnel sides

Funnel front

Funnel corner supports

Ramp

Vertical ramp support

Base

Ramp cleat

PROJECT OVERVIEW

Approximate materials cost: $20

Time to build: 2-4 hours

Level of complexity: Moderate to difficult

SUGGESTED TOOLS

- Circular saw
- Drill
- Clamps

RECOMMENDED MATERIALS

Less than half a sheet of $3/8$" or $1/2$" plywood

One 2×4 or 2×6, about 26" long

Five scraps of 2×2 lumber, each 6"–10" long

Thirty-six $1\frac{1}{4}$"–$1\frac{1}{2}$" screws

TRIANGLE GEOMETRY

If you know two of the angles and the length of one side of a triangle, you can calculate the remaining angle and the lengths of the other two sides.

The sum of the three internal angles for every triangle is 180 degrees. So if you know that one of your angles is 90 degrees and another angle is 45 degrees, the third angle must also be 45 degrees because 90+45+45=180.

This is the case with the ramp in this project. The ramp itself will sit at a 45-degree angle, and the vertical support will connect to the base at a right angle (step 4 on page 117). Therefore, the remaining angle must also be 45 degrees.

Triangles with two matching angles are known as isosceles triangles, and the lengths of the sides opposite those angles are always equal. If the top of the ramp is 32 inches above the ground (side a), then side b must also be 32 inches.

You can calculate the length of side c (in this case the ramp itself) using basic algebra and the Pythagorean theorem: $a^2+b^2=c^2$. The Pythagorean theorem works for any right triangle where c is the length of the hypotenuse (the side opposite the right angle) and a and b are the lengths of the other two sides. In this case, $32^2+32^2=2,048$. The square root of 2,048 is slightly more than 45. For our purposes, it's fine to round to 45.

Many other designs in this book are also right triangles. For more about using basic geometry, algebra, and trigonometry in tool designs, see page 184.

HOW TO BUILD IT

1 **Cut the ramp.** Measure your harvest container and cut the plywood for the ramp slightly wider than that measurement. In my case, our harvest containers are 15" wide, so I made the ramp about 16". The ramp will sit at a 45-degree angle. To determine the length of the ramp, first decide how high off the ground you want the top edge of the ramp to be; this should be a comfortable dumping height. For example, I want the top of my ramp 32" above the ground, so I make the ramp 45" long. To understand how to calculate your dimensions, see Triangle Geometry at left.

2 **Cut the base.** Cut a piece of plywood roughly the same width and length as the ramp—in this case 16" × 45". It's fine if it's a bit shorter or narrower, and if you're short on plywood you can modify the base design and use other wood.

3 **Cut the vertical ramp support and cleat.** For the vertical support, cut a 2×4 or 2×6 with one square end and one end at a 45-degree angle, with the long edge measuring 26". For the cleat, cut a piece of 2×2 with square ends, roughly the width of your ramp—in this case 16".

4 **Assemble the ramp.** Secure the vertical ramp support piece to the center of one end of the base with two screws driven through the plywood base and into the square end of the vertical support. The 26" edge of

the vertical support should be flush with the 16" edge of the base piece. Rest the ramp on the base piece, flush against the 45-degree angle of the vertical support, and drive two screws through the plywood into the vertical support to hold it in place.

Finally, slide the cleat under the bottom end of the ramp. Fasten it by driving two screws through the ramp and into the cleat, then drive two more screws through the bottom of the base and into the cleat. The cleat won't sit flush against the ramp at the base, so just aim to get the screws close to the edge where the cleat contacts the ramp. Make sure to get all screw heads flush with the plywood surface or to sink them slightly to prevent them from damaging your produce as it rolls down the ramp or from damaging your floor (see Getting Screw Heads Flush on page 112).

5 Cut the funnel front and corner supports. Cut a piece of plywood about 16" wide and between 8" and 12" tall for the front of the funnel. In addition, cut four pieces of 2×2 6"–10" long. These will give you something to screw into and hold the funnel pieces together; they don't all need to be precisely the same length.

6 Cut the funnel sides. You want the funnel opening to be slightly smaller than the mouth of your mesh bag to ensure that your bag fits well over the opening. Measure the mouth of the bag; in our case, the bag is 18" wide, so its opening is about 36" all the way around. Making the opening of the funnel 32" around, or 8" square, leaves enough space to fit the bag mouth outside the funnel opening while still being large enough to let most potatoes and onions roll through unimpeded.

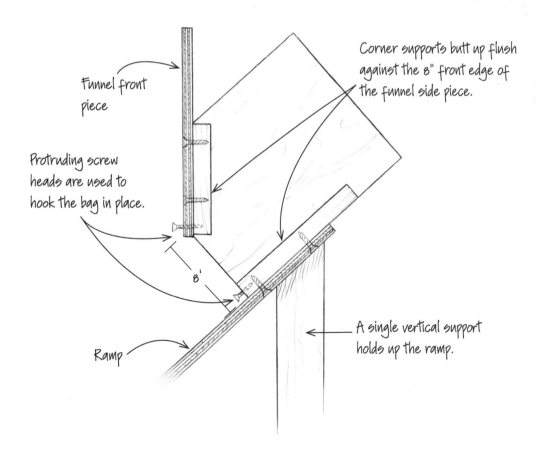

Funnel front piece

Corner supports butt up flush against the 8" front edge of the funnel side piece.

Protruding screw heads are used to hook the bag in place.

8"

A single vertical support holds up the ramp.

Ramp

Start by cutting two pieces of plywood approximately 20" × 16". Leaving 8" on the 16" edge, cut an angle of about 45 degrees off the corner. The exact angle isn't important, but the two side pieces should match. To make it easier to match them, clamp the two pieces of plywood together and cut both pieces at the same time.

7 Attach the funnel corner supports. Attach two 2×2 supports to the outside face of each funnel side piece, using two screws per support. One support runs along the angled edge at the front corner; the other runs along the long edge. Both supports should be flush with the front (8") edge of the side piece. Drive the screws through the plywood and into the supports. Each side should be a mirror image of the other, with the corner supports on opposite faces of the two pieces.

8 Install the funnel sides. Draw a line across the ramp, parallel to and 12" from the top edge. Mark the center of that line, then measure 4" to either side of the center and make a mark as shown. Position one funnel side piece on the ramp with the 2×2 support pieces facing the outside edge of the ramp, the 8" edge of the plywood aligned with one of the 4" marks, and the long plywood edge extending up past the top corner of the ramp to prevent spillage when you pour vegetables into the funnel. Clamp the side piece in place and secure it with two screws going through the ramp plywood and into the 2×2 support. Repeat for the other side.

9 Mount the funnel front. Place the plywood front piece against the angled front edges of the funnel sides, with the bottom edge flush with the bottom of the funnel

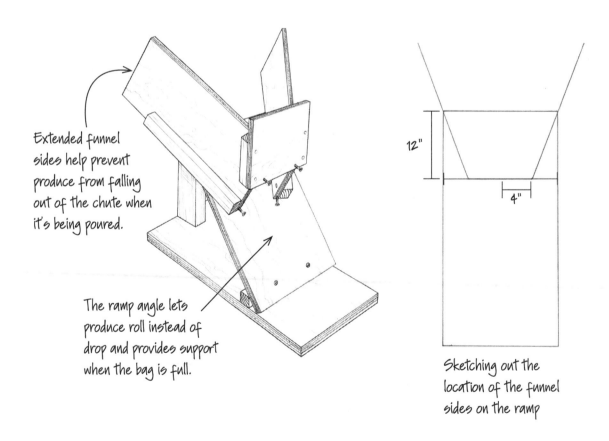

Extended funnel sides help prevent produce from falling out of the chute when it's being poured.

The ramp angle lets produce roll instead of drop and provides support when the bag is full.

12"

4"

Sketching out the location of the funnel sides on the ramp

opening. Secure the front piece with two screws going into the 2×2 support on each side. Note that the front piece will not sit exactly flat against the support pieces and there will be a small gap because the side pieces are slightly angled. Functionally this is not a problem.

10 **Add hook screws to the funnel supports.** Into the end grain of each 2×2 funnel support attached to the ramp, drive one screw about $\frac{1}{8}$" above the ramp, close to the corners of the opening. Leave the screw heads sticking out about $\frac{1}{4}$", just far enough so you can slide the bag behind them. These

screws will hold the rear edge of the bag close enough to the ramp that onions or potatoes rolling down the ramp won't slide behind the bag opening.

11 **Add hook screws to the funnel front.** Hook one of your bags over the screws you attached in step 10, then pull the bag open across the funnel opening. Mark where the opposite corners of the bag opening hit the top of the funnel, leaving just enough slack so you can hook the bag to all four screws at the same time. Place one screw at each of these marks, again leaving the head sticking out about $\frac{1}{4}$".

USING THE ONION BAG FILLING STAND

Hook a bag to the four screw heads at the corners of the funnel, starting with the two screws closest to the ramp. Putting a bit of tension between the connections on the ramp side and between the ramp and the front of the bag prevents the bag from coming loose and makes it lie as flat as possible against the ramp.

Once the bag is fully hooked on, slowly pour your crop into the top of the funnel. We generally weigh out 25-pound totes of potatoes or onions, then pour two of those through the funnel to fill one 50-pound bag. This works well because our scale has a 40-pound capacity and because 25-pound totes are much easier to lift and pour than 50-pound totes. The design allows most of the weight to sit on the ground, with some support from the ramp, so there isn't too much tension pulling down on the bag where it is hooked to the funnel.

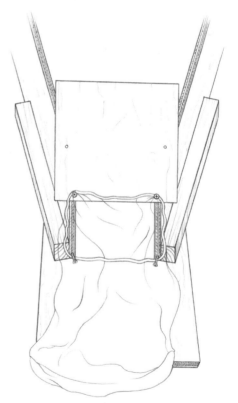

It's important to place the four screws so the bag opening fits tightly and doesn't slide off while you're filling it.

SIMPLE SPRAY TABLE

After harvest, many crops need to be washed, trimmed, and inspected for quality before being packed and distributed. The Simple Spray Table provides a stable surface at a comfortable working height. This one has a slatted top that allows water and mud to drain away without letting crops fall through or get stuck. The surface is smooth enough that it won't damage the crops, and as an added convenience, crops or containers can slide from one end of the table to the other as they are cleaned and inspected.

You can add a low stand to serve as a dunnage rack (intended to lift materials up off the ground) at either, or both, ends of the table. I use a rack on one end to stack crates of dirty produce that need to be cleaned and sorted and another rack at the other end for clean produce. By keeping the produce off the ground, the rack is good for both food safety and worker safety, since it requires less bending. The height of the racks puts the crates just below the edge of the tabletop, letting us roll or slide cleaned and sorted produce off the table and into the crate without any lifting.

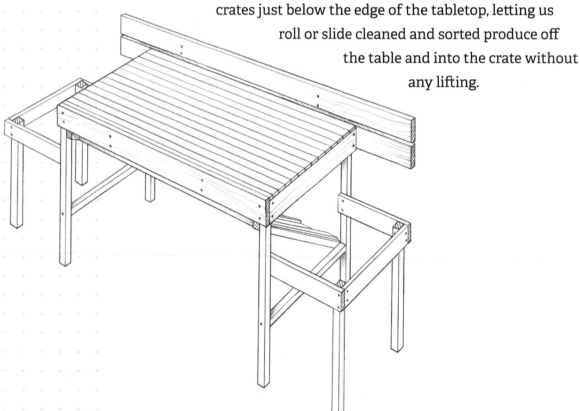

PROJECT OVERVIEW

Approximate materials cost: $60

Time to build: 4 hours

Level of complexity: Moderate

SUGGESTED TOOLS

- Saw
- Drill
- Speed square
- Tin snips
- File
- Pneumatic stapler or basic hammer

RECOMMENDED MATERIALS

Four 96" lengths of 2×2	2" screws
Three 96" lengths of 1×6	2½" screws
Fourteen 48" lengths of lath, typically $5/16$" × $1\frac{1}{2}$"	Gasketed metal roofing screws
Ribbed metal roofing, approximately 36" × 48"	1" staples (¼" crown or larger) or 4d (1½") common nails

NOTE: For each dunnage rack you include, you'll need an extra 48" (or more) of 2×2 and 1×6.

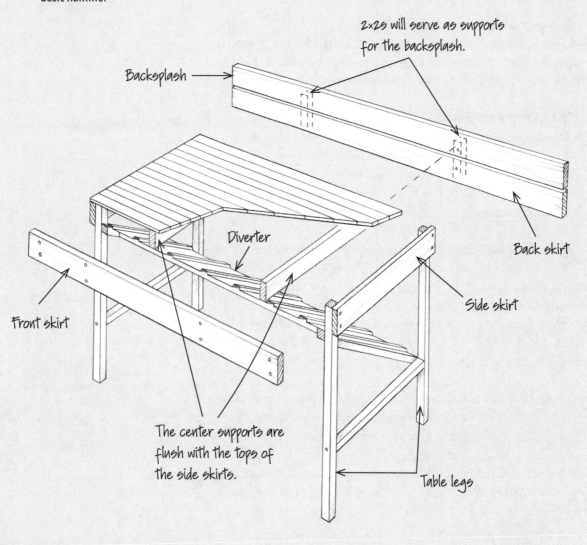

2×2s will serve as supports for the backsplash.

Backsplash

Back skirt

Diverter

Front skirt

Side skirt

The center supports are flush with the tops of the side skirts.

Table legs

HOW TO BUILD IT

1 **Cut the table legs, skirts, and center supports.** Cut four 38" legs, each from a different 96" piece of 2×2. You can adjust the length to customize the table height, but keep in mind that the lath will add about $5/16$". Cut one 48" length from one of the 1×6s for the front skirt, then cut the remaining piece of that board exactly in half for the side skirts (making them approximately 24" each). From two separate 1×6 boards, cut a length that exactly matches the side skirts to get two center supports. Use the remaining length of both of these 1×6 boards for the back skirt and backsplash (making them approximately 72" each, but the exact length is not critical).

2 **Join the legs and side skirts.** Position each leg so it is flush with the top edge and end of a side skirt. Join the pieces by driving two 2" screws through the side skirt and into the leg. Keep the screws about ¾" from the edges of the side skirts.

3 **Install the front skirt.** Attach the front skirt to the legs with two 2" screws per connection. The ends of the front skirt should be flush with the outside faces of the side skirts. The top edge of the front skirt should be ½"–¾" above the top edges of the side skirts. This creates the small lip at the front edge of the table that keeps produce from rolling off easily.

4 **Add the back skirt.** Center the back skirt over the side skirts so that it extends roughly 12" beyond both side skirts. Fasten the back skirt to the legs with two screws at each connection, making sure that everything is square and that the outside edges of the side skirts are 48" apart. Positioning the skirt slightly higher than the legs is less critical in the back than it is in the front, but I prefer to have the back match the front.

5 **Add the center supports.** Place the two 1×6 center supports about one-third of the way (approximately 16") from each end so that they span the gap between the front and back skirts and are flush with the tops of the side skirts (not flush with the front and back skirts). Attach by screwing through the front and back skirts and into the end grain of the center supports.

6 Add the side stretchers. Measure the distance between the inside edges of the front and back legs. Cut two lengths of 2×2, one from each of the pieces remaining from cutting the legs, to this dimension; it should be about 21". Fit these stretchers between the front and back legs on either side of the frame, about 6" above the bottoms of the legs, and fasten them with a single 2½" screw through each leg and into the end grain of the stretcher.

7 Build the diverter frame. The diverter frame is made of 2×2s and supports the piece of metal roofing material while providing diagonal bracing for the legs. For the diagonal braces, cut two lengths from the same 2×2s from which you cut the stretchers in step 6. Cut one end of each diagonal brace at a 45-degree angle, using a speed square to mark the angle. Line up the angled end of one brace flush with the front of one front leg and mark a line where it crosses the back of the back leg. Cut along that line. Cut the second piece to match. These pieces will end up being about 36" long from tip to tip.

Cut two stretchers, 40½" each, from the remaining 2×2s; both of these are square-cut. Check the 40½" length before cutting; it should match the distance between the insides of the legs minus the width of the two diagonal braces. Position the stretchers between the braces, about 2" from their ends. Assemble the frame with 2½" screws driven through the diagonal braces and into the ends of the stretchers.

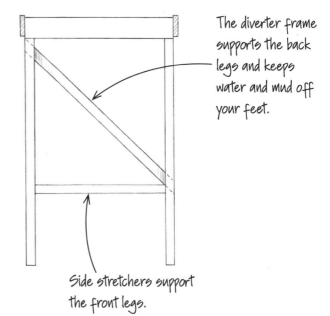

The diverter frame supports the back legs and keeps water and mud off your feet.

Side stretchers support the front legs.

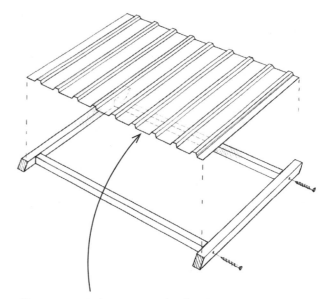

The corrugated roofing on the diverter frame redirects water and mud and also helps stiffen the whole table significantly.

8 Add the diverter panel. Cut the metal roofing to match the length of the diagonal braces of the diverter frame, using tin snips (see the tip below). The ridges in the corrugation should run parallel to the diagonal bracing (so the water flows downward). Fasten the metal to the frame with gasketed metal roofing screws. Use one screw at each corner and three or four screws along each edge.

9 Install the diverter. The top edge of the diverter frame needs to sit near the bottom edge of the front skirt. It can sit either just at the bottom of the skirt (in which case it will be hidden from view) or an inch or two below the bottom of the skirt (in which case it will be easier to clean the diverter, but the metal edge will be somewhat exposed). The frame should angle down to the back legs at a 45-degree angle. Fasten the diverter frame to all four table legs with one 2½" screw per leg.

TIP CUTTING METAL

Be careful when working with sheet metal roofing, as the edges are sharp and it's easy to cut yourself. It's a good idea to wear leather gloves and use a file to dull any raised burrs on the edges. Compound snips, which work well but force you to go slowly, usually have color-coded handles; you will want yellow-handled snips for making straight cuts.

You probably won't find a piece of roofing that's exactly the width you need, but you may be able to overlap two pieces and not have to cut the edges. If you do have to cut the edges, cut the metal just slightly undersize from side to side so it doesn't stick out anywhere once it's attached to the wood frame.

You do want the metal to extend to the ends of the diagonal braces, though. The front skirt will cover the top edge of the diverter, and you can set up your table so that no one can walk behind the diverter, where the water and mud pour out. If you want the water to pour into a diversion gutter, leave the metal even longer so that it will overhang the gutter slightly (or you can add a piece of flashing later).

---

10 Make the tabletop surface. Place the lath strips across the tabletop, leaving a gap of about ¼" between neighboring strips and between the first and last strips and the front and back skirts. (Most speed squares are about ¼" thick, and using one as a spacer makes it easy to gauge this distance.) The dimensions of different batches of lath can vary quite a bit, but the exact spacing between the strips isn't critical as long as it's somewhere between ⅛" and ⅜". Lay out the strips and discard any that have significant knots or that are already starting to crack or warp. Attach the strips to the side skirts and center supports with a pneumatic stapler, if you have one, or nail the strips with 4d common nails. To prevent the lath and 1× from splitting, you may need to drill pilot holes for the nails.

11 **Build the backsplash.** Cut two 10" lengths of 2×2 for the backsplash supports. Position each support against the back skirt so it extends 5" above the top edge of the back skirt. It is easiest to place the supports just outside the table legs, but the exact placement is not critical. Fasten the supports through the front of the skirt, using 2" screws. Position the 1×6 backsplash against the supports so its ends are aligned with the ends of the back skirt and there is a gap of about ¼" between the top of the back skirt and bottom of the backsplash. This will let a little water get through for cleaning. Screw the backsplash to the supports.

12 **Build the optional dunnage rack(s).** This simple addition works with harvest containers that have a base at least 20½" long. When placed on a rack, the harvest container will sit just below the edge of the table. To determine the length of the legs for your rack(s), subtract the height of your harvest container from the height of the tabletop. For each rack, cut two pieces of 2×2 just shy of this measurement.

For each rack, cut two pieces of 1×6 about as long as your harvest containers are wide. Position one leg flush with the end and top edge of one 1× piece. Fasten into place with two screws. Repeat with the other leg, making a mirrored copy to get one front and one back. Screw the other ends of the 1× pieces to the inside of the main table legs, making sure the 1x pieces are level. The short legs should face each other and will be on the opposite side of the 1× from the table leg to which they are connected.

Measure the distance between the outside faces of the two 1x pieces, and cut a length of 1×4 or 1×6 to that dimension; it should be about 20½". Screw this piece to the short legs to hold the end together. No additional bracing is necessary, as the table is stiff enough to hold everything in place, and the short legs don't allow much torque on the connections.

On the dunnage rack, the top of the harvest container sits just below the table surface, making it easy to slide produce off the table and into the container.

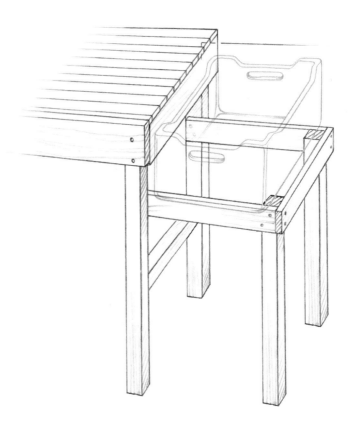

USING THE SPRAY TABLE

I really like using this table, and it's fairly easy to build. If you don't set up your table on a hard slab, you'll need to add broad feet to keep it from sinking into gravel or mud. You might consider adding 2×4 blocks to the bottoms of the legs anyway as sacrificial pieces that will eventually rot but can be replaced easily.

The biggest problem with this table is that it can be tricky to make sure that every nook and cranny is cleaned after each use. A strong, thorough spray-down with a hose should get rid of most things, but sometimes you may need to scrub the table with a nylon brush to remove stubborn, stringy bits that get into the cracks.

Generally wood is not considered food safe because it is porous and doesn't react well with sanitizers. I've talked with many experts, though, and the only scientific research on food safety and wood that any of them has pointed me to suggests that wood is probably at least as safe as, and possibly safer than, plastic surfaces, which are commonly considered food safe. In multiple studies, Dean O. Cliver, PhD, concluded that wooden cutting boards are probably at least as safe as plastic cutting boards, and in some respects safer. If you're concerned about building this table out of wood, though, you could substitute high-density polyethylene plastic strips for the lath. I've also seen similar designs that use perforated stainless steel.

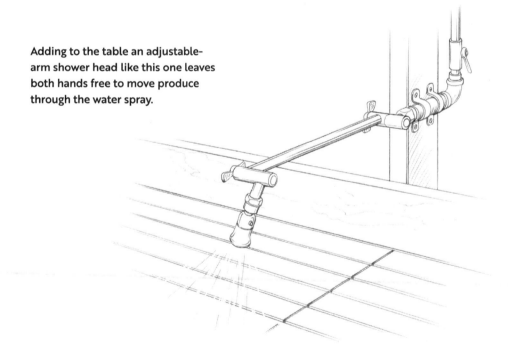

Adding to the table an adjustable-arm shower head like this one leaves both hands free to move produce through the water spray.

DESIGN NOTES

I designed the tabletop of my spray table with spaced wood lath, the same material I used for the Simple Seedling Bench on page 26 and that I describe in the note on page 30. My table design also incorporates a backsplash to keep crops from rolling off the back of the table as I work, plus a slight lip at the front. I find that a lip of ¼ inch to ½ inch is enough to keep most things from rolling off, while still allowing easy access to load bins and crops onto the table.

Below the tabletop, a diverter panel made with corrugated metal roofing sends water and soil away from the user's legs and feet. This makes it much more comfortable to work at the table, especially in the cold months, and it helps prevent a giant mud puddle if you're working on a dirt floor.

The table dimensions—2 feet deep by 4 feet wide—let me work comfortably across the whole table. I can reach about 2 feet in front of me without having to strain at all, and I don't have to move very much from side to side to reach the edges.

Over the years I've made a number of variations on this table, but they all include the features shown in this project. To make the table lighter, so that one person can move it (without any wheels), you can build the table and dunnage racks separately.

At my farm, I created a built-in version of this table in a small pole shed and incorporated extra-wide dunnage racks that hold multiple bins. The whole thing is fixed to the larger structure. With that construction, I've wished that I ran the roofing piece across the entire width of the dunnage racks and table, which could have been accomplished with a single long piece of roofing angled to drain to one side instead of to the back.

I've also made a few add-ons to my basic design, including a gutter along the back of the table to move the runoff water away from the table. In addition, I've incorporated an adjustable-arm showerhead, which lets me spray off produce without having to use one hand to operate the sprayer. On my farm we have very limited water so the showerhead I use is very low flow, which saves water, but doesn't save much time. It's likely that a design that used more water would also be more effective at cleaning produce.

In the future I'd like to incorporate a Dramm thumb valve or a foot-operated valve to turn the spray water on and off easily. The Dramm valve would require a little modification, but if it were mounted on the front skirt, it could be turned on by simply leaning against it. Both types of valves are spring loaded, so they would turn themselves off automatically.

The table design—including its height, width, and depth—takes into account human dimensions.

MINI BARREL WASHER

My friend Danny washes a lot of roots on his small farm in Ridgefield, Washington. After years of washing by hand on a spray table, he asked me what I knew about barrel washers. A barrel washer is essentially a long tube, open on both ends, with a pipe hanging inside that sprays water. The barrel is positioned horizontally and at a slight tilt. Roots are fed into the higher end as the barrel spins, which causes the roots to tumble down the barrel tube and out the lower end. All the while, the pipe sprays water to wash the roots.

While most commercial barrel washers are very large and/or very expensive and use high volumes of water, Danny needed a smaller version that would fit into his small packing shed and wouldn't use a lot of water. I sketched out a design and built this relatively compact prototype version for Danny that would use

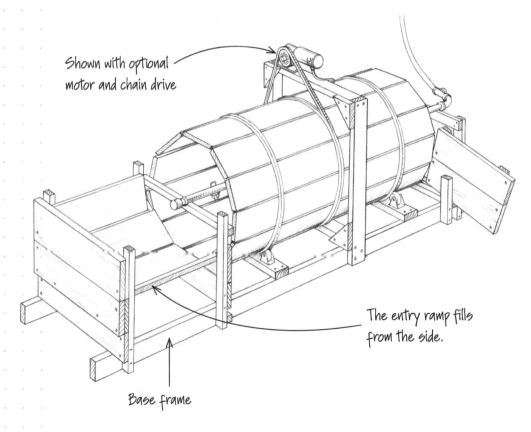

Shown with optional motor and chain drive

The entry ramp fills from the side.

Base frame

less water by having a shorter barrel and fewer sprayers, with the spray nozzles focused more directly on the crop.

The barrel is made with fence boards screwed to two aluminum bicycle rims. The rims ride on two pairs of caster wheels mounted onto a base frame. Gaps between the boards in the barrel allow water and dirt to drain away before the cleaned roots exit the washer. The barrel can be turned by hand or by an electric motor (Danny eventually added a motor drive; see Using the Mini Barrel Washer on page 137).

To compensate for the relatively short barrel length, this washer includes a gate at the lower end of the barrel to hold the crop under the spray for longer when needed. To make it easier to load roots into the barrel in a compact space, I added a funnel at the top end of the barrel, where a harvest crate can rest on its edge as it is tipped up and emptied from the side. I included a similar ramp at the exit to feed the outflow into a clean crate.

Because this was a prototype, built mostly with salvaged materials, I'll walk you through the process to decide on dimensions as you build your own version.

NOTE: When sourcing materials for this project, try to find a pair of matching bike rims, or at least rims that have the same inner diameters and outer diameters, then find four fixed (not rotating) casters that fit nicely inside the rims. The project as shown uses standard single-wall 26-inch rims.

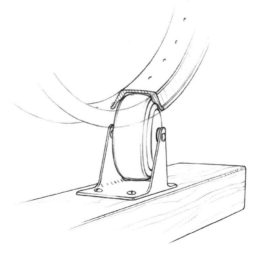

PROJECT OVERVIEW

Approximate materials cost:
$100–$200

Time to build: 8–12 hours

Level of complexity: High

SUGGESTED TOOLS

- Saw
- Drill
- Screwdriver
- Framing square

RECOMMENDED MATERIALS

Two 26"-diameter bicycle rims

Fourteen lengths of cedar fence board, about $5/8$" thick × $5\frac{1}{2}$" wide

2×4s, totaling approximately 20'

1×2s, totaling approximately 25' (I used wooden contractor stakes left over from staking peppers)

Four fixed casters (and mounting screws) with wheels narrow enough to fit inside your rims

$\frac{1}{2}$" sheet metal screws (to connect rims to fence boards)

$1\frac{1}{4}$" screws (to connect stakes and fence boards)

$2\frac{1}{2}$" screws (to connect 2×4s)

Three K-Ball clip-on nozzles (or similar)

Schedule 40 PVC pipe, sized to match the nozzle requirement and barrel length

PVC elbow, sized to match the pipe, slip × $\frac{3}{4}$" female pipe thread (FPT), or equivalent adapters to transition to FPT

$\frac{3}{4}$" × 6" PVC threaded pipe nipple

$\frac{3}{4}$" PVC ball valve, FPT × FPT

$\frac{3}{4}$" male pipe thread (MPT) × female hose thread adapter

PVC cap, sized to match the pipe

Thread tape, PVC primer, and glue

Synthetic baling twine or plumber's strapping

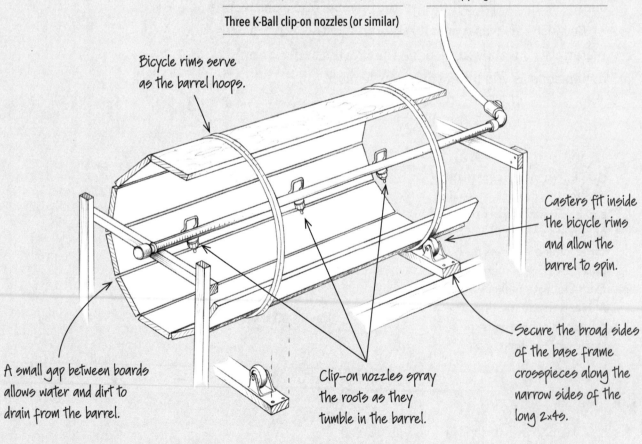

Bicycle rims serve as the barrel hoops.

Casters fit inside the bicycle rims and allow the barrel to spin.

A small gap between boards allows water and dirt to drain from the barrel.

Clip-on nozzles spray the roots as they tumble in the barrel.

Secure the broad sides of the base frame crosspieces along the narrow sides of the long 2×4s.

HOW TO BUILD IT

1 **Build the barrel.** Measure the inside diameter of your bicycle rims and multiply that length by π (3.14) to get the inner circumference. Divide that number by the width of your fence boards to get an estimate of how many you'll need. In my case I needed about 11 fence boards, which I cut to 48" lengths.

Mark the outside faces of the boards at about 9" from each end. Use these marks to guide the rim placement so that the ends of the boards create an even barrel and the rims are perpendicular to the boards. Fasten each board to the insides of the rims, using two ½" sheet metal screws per rim and going through the rim and into the board. Place the screws about ¾" from each edge of the board.

Leave a gap of about ⅜" between neighboring boards. The gaps don't all have to match, but it's best if they're not too narrow (which will restrict water and dirt from flowing out) or too wide (which might cause small roots to get stuck or fall through). If your last board doesn't fit well, rip it to an appropriate width using a table saw.

2 **Build the base frame.** Cut two 2×4s that total the length of the barrel plus the length of the exit ramp and the length of the full entry ramp assembly. In my case, the barrel was 48", I made the exit ramp from a single fence board, and the entry ramp was a bit longer than the width of a harvest tote, so the full length of the base frame was about 72". Alternatively, you can leave the 2×4s long and trim them after you have completed the other steps.

Cut two additional 2×4s a bit longer than the diameter of your barrel. In my case, this was about 24". Place the broad sides of these crosspieces along the narrow sides of the long 2×4s. The space between the centers of the crosspieces should be the same as the distance between the centers of the hoops on the barrel. Make sure everything is square, and use two 2½" screws through each crosspiece joint to hold it all together.

TIP FITTING THE RIMS

Bicycle rims have holes in them for spokes, but those may not line up well with where you'll want a screw, so plan on drilling more holes as needed in the rims for screws.

You can vary the length of your barrel boards as desired (to make your barrel longer or shorter), but keep in mind that you may need to adjust the locations of the rims, and a much longer barrel may need additional rims for adequate support; see Design Notes on page 138. If you adjust the dimensions of your barrel, you will also need to adjust the dimensions of the barrel frame.

3 **Mount the casters**. Your casters will sit on the crosspieces of the base frame. To figure out the spacing between the casters, hold two casters over the barrel, with their wheels sitting in the groove of one of the bicycle rims. Set a straight board or measuring stick across the top of the two casters. Then, holding one caster and the board in place, slide the other caster along the length of the board, keeping the wheels in the groove of the bicycle rim. When the casters are close together the board will be as far from the rim as possible, and as you slide the casters apart the board will get closer to the rim. You want to stop when the rim is an inch or less from the board. If the casters are too far apart, the barrel might drag on the frame under weight and as the wheels wear. If the casters are too close together, the barrel will be less stable.

Measure the space between the casters, then transfer that measurement, centered, to the crosspieces on the base frame. My casters were about 18" apart, which put each about 3" in from the ends of the crosspiece. Predrill the holes for the casters, then fasten the casters in place with the mounting screws. The casters should be lined up with each other and with the rims on the barrel. Sit the barrel on the casters. You should be able to spin the barrel with little or no resistance.

4 **Prepare the exit ramp.** In the steps that follow, you can take your measurements directly from the barrel and the base frame you've just assembled.

Cut a piece of fence board for the ramp the same width as the 2×4 base frame.

Cut two lengths of 1×2 just long enough so that, when they are placed vertically on the outside edge of the frame, they extend up just shy of the barrel. These are the inside vertical support pieces. Carefully predrill two holes for

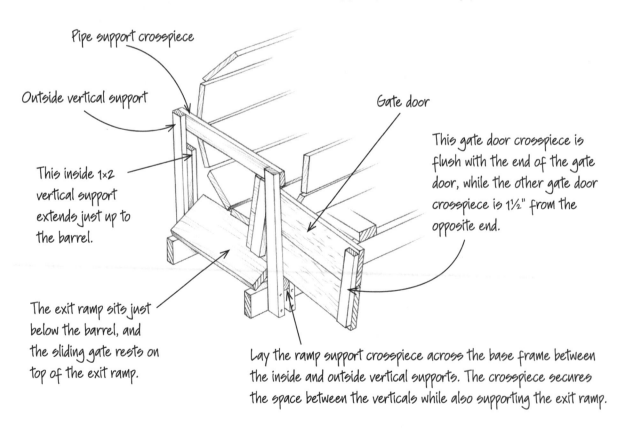

Pipe support crosspiece

Outside vertical support

Gate door

This inside 1×2 vertical support extends just up to the barrel.

This gate door crosspiece is flush with the end of the gate door, while the other gate door crosspiece is 1½" from the opposite end.

The exit ramp sits just below the barrel, and the sliding gate rests on top of the exit ramp.

Lay the ramp support crosspiece across the base frame between the inside and outside vertical supports. The crosspiece secures the space between the verticals while also supporting the exit ramp.

screws through the 1× face of each piece. Secure these two pieces vertically to the outside of the frame, so their broad faces are just ⅛" to ¼" in front of the barrel and with their narrow edges are against the 2× rail. These pieces will keep the gate from scraping on the edge of the barrel.

Cut two 1×2 pieces that span the distance between the outside edges of the two vertical pieces you just installed. In my case, this was about 24" + 1½" + 1½", or 27" total (the width of the base frame plus two widths of the 2× material). Put one of these pieces aside; you'll install it as the pipe support crosspiece in step 6. Lay the other piece across the base frame just in front of the inside vertical supports, with the narrow edge against the 2×4 rails. This is the ramp support crosspiece.

Lay the fence-board ramp piece so that it angles down from the top edge of this crosspiece to the ends of the 2×4 rails on the base frame. If the ramp now hits the edge of the barrel, remove the ramp support crosspiece and use a table saw to rip it, making it a little narrower.

Wait to screw all of the exit ramp pieces in place until you complete the gate assembly in step 6.

5 **Construct the gate.** Cut two pieces of fence board 1½" longer than the pipe support crosspiece that you cut in step 4. Cut two pieces of 1×2 the width of the two fence boards. Use 1¼" screws to secure one of these 1×2 lengths across the two fence boards, flush with one end. Screw the other piece of 1×2 across the fence boards 1½" in from the opposite end to complete the gate door.

Cut six more lengths of 1×2. At a minimum these need to be long enough to extend from the bottom of the 2×4 base frame to 1¾" above the gate when it sits on top of the exit ramp. Two of these pieces will become the outside vertical supports. The other four will be set aside for use in building the entry ramp in step 7. Adding up to 6" of extra length can be helpful, especially for the verticals you'll use for the entry ramp.

6 **Complete the gate assembly.** Using 1¼" screws, attach the two 1×2 outside verticals you cut in step 5 to the base frame in front of the 1×2 ramp support crosspiece. Pin the crosspiece in place with a screw through the verticals and through the crosspiece on both ends. The crosspiece should now be sandwiched between the inside and outside verticals on either side of the base frame. At this point you can secure the ramp in place with four screws, one through the ramp into each of the 2×4 frame pieces, and one at each of the top corners of the ramp into the crosspiece.

Slide the gate door down between the inside and outside verticals and let it sit on top of the ramp. The 1×2 crosspiece that is flush with one end of the door should sit outside the verticals on the side of the assembly where you'll be standing when washing roots. This crosspiece will act as a handle for pulling the gate open. The other end of the gate should sit between the verticals on the other side of the base frame. Screw the pipe support crosspiece you set aside in step 4 across the tops of the two outside verticals to keep them spaced properly and to hold the gate down. (Eventually this crosspiece will also hold up one end of the pipe for the spray heads.)

Now the gate door should be sitting on the ramp, prevented from moving forward or backward by the inside and outside verticals but able to slide to the side to let roots out from the end of the barrel. The gate isn't able to be easily removed, which keeps it from getting lost.

7 **Prepare the entry ramp parts.** Cut four fence boards the width of your base frame plus the width of two 1×2s. Again, in my case, this was about 27" total. These are crosspieces that will hold together the 1×2 verticals, provide support for the ramp, and keep roots from tumbling out.

Cut four more pieces of fence board, each with one end that is 45 degrees and the other end square. The long edge of these four pieces should be 10". (This assumes your fence boards are 5½" wide and ⅝" thick, and that the previous cut was 27"; you may need to adjust the length of the long edge if your fence board dimensions are different.) These boards will form the ramp cradle. Finally, cut four more fence boards that match the width of your harvest containers plus 5". Our harvest containers are 15" wide, so I cut these last four boards to 20" lengths. These boards will form the entry ramp itself.

8 **Build the entry ramp supports.** Position two of the four extra 1×2s that you cut in step 5 as the inside vertical supports, setting them on the outsides of the base frame rails, so that a fence board fits between them and the face of the barrel, with about ¼" clearance to prevent rubbing. Screw these verticals to the 2×4 base frame rails in the same way you did the verticals in step 4, orienting them so the narrow side contacts the 2×4 frame, and carefully predrilling for two 2½" screws per vertical.

Set one of the fence-board crosspieces that you cut in step 7 across the base frame rails so that it sits between the verticals and the barrel. Its top edge should sit at least 1" above the inside of the barrel. Screw it into place using 1¼" screws.

Position the second set of 1×2s you cut in step 5 as the outside verticals. The distance between the outside and inside verticals should be the width of your harvest tote plus about 1".

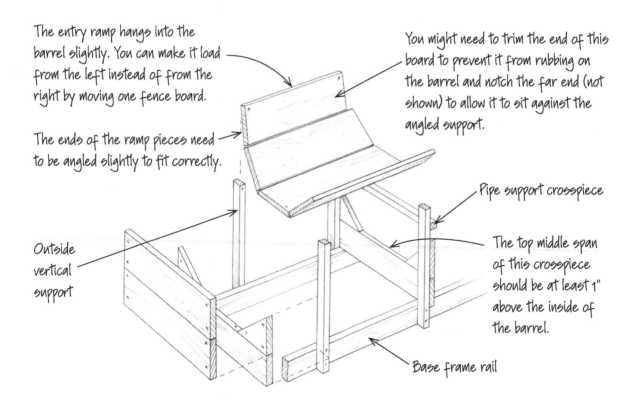

The entry ramp hangs into the barrel slightly. You can make it load from the left instead of from the right by moving one fence board.

The ends of the ramp pieces need to be angled slightly to fit correctly.

You might need to trim the end of this board to prevent it from rubbing on the barrel and notch the far end (not shown) to allow it to sit against the angled support.

Pipe support crosspiece

The top middle span of this crosspiece should be at least 1" above the inside of the barrel.

Outside vertical support

Base frame rail

This spacing lets the harvest container slide between the verticals but doesn't provide so much room that roots fall out. Position another fence-board crosspiece across the outside face of the outside verticals an inch or two higher than the crosspiece on the inside verticals, then screw it in place.

Set one pair of the ramp cradle pieces that you cut in step 5, with 45-degree ends, on top of each of the two crosspieces you just attached. The square ends of these cradle boards should sit flush with the outside edge of the verticals, and the long edges will sit on the crosspieces below. Screw the front set of the cradle boards directly to the inside verticals using 1¼" screws. On the back set, place a full-length fence-board crosspiece behind the two cradle pieces, screwing through both boards and into the outside verticals. Screw the remaining crosspiece to the outside verticals above the last crosspiece, creating a high back to keep the roots from tumbling out.

9 **Assemble the entry ramp.** Lay one of the longer ramp boards you cut in step 7 across the inside and outside crosspieces, centered between the angled cradle boards. It should extend an inch or two into the barrel, with enough clearance to allow the barrel to rotate freely. Screw the ramp board down to both crosspieces using 1¼ screws. Then screw down two more ramp boards, one on each of the angled cradle boards, leaving a slight gap between the boards to make cleaning easier. Attach the last ramp board above the angled cradle board opposite the side from which you want to pour. Trim the ends of the boards sitting on the angled cradle boards so that they sit flush against the outside crosspiece. This is easiest to measure when putting them in place. You may also need to notch the last piece so it sits properly on the angled support, which is partially obstructed by the board above it.

10 **Build the pipe support.** Cut another 1×2 to screw across the tops of the inside verticals of the entry ramp. Using 1¼" screws, fasten it in place at the same height as the pipe support crosspiece on the exit gate end.

11 **Dry-fit the water pipe.** Cut a length of PVC pipe long enough to span the distance between the pipe support crosspieces at the entrance and exit. On whichever end is more convenient to feed the pipe from, place an elbow that transitions to a ¾" nipple and then the ball valve. Transition into a hose thread adapter so that you can plug in a garden hose (if that is your water source). Cap the other end of the pipe. Remember to add thread tape to the pipe threads (see page 93).

12 **Drill holes for the nozzles.** In the bottom of the pipe, carefully drill three holes sized for your nozzles (mine were ⁹⁄₁₆"). It is easy to crack the pipe, especially when using a twist bit or a spade bit, so clamp the pipe securely against something that will prevent it from twisting as you drill. Preferably, use a hole saw or Forstner bit to apply light pressure and high speed to make the hole. If you have a drill press this is a good place to use it. Clip the nozzles to the pipe.

13 **Mount the pipe in the barrel.** Tie the pipe in place with synthetic baling twine or use a small amount of plumber's strapping to hold it down with screws. Position the nozzles so they will be as close to the roots as possible without getting in the way as the roots move through the barrel. Adjust the nozzles to spray toward the roots. Once you are satisfied with the pipe setup, use PVC primer and glue to secure the cap and elbow and prevent blowouts.

LOOKING CLOSER

FLOW RATE CALCULATIONS

Instead of buying nozzles, you might be tempted to drill holes directly into the water pipe or to make very shallow 45-degree angled cuts with a table saw that would just nick the interior diameter of the pipe and create small slits. This is worth experimenting with, but if you do, keep in mind the following back-of-the-envelope calculations, the principles behind them, and the potential complication of trying to get the water to come out in the direction and spray pattern you want.

You can calculate the exit speed of water from a hole by knowing the water pressure. The speed combined with the size of the hole will give you the flow rate coming out of the hole in a volume per time, such as gallons per minute (gpm). Ultimately, the shape of the hole, and more specifically the lead-in and exit geometries, also will have an impact on the speed and shape of the water coming out. (For example, the water can exit a hole in the shape of a fan, a cone, or a narrow stream.)

Running those numbers for a small, $\frac{1}{16}$-inch-diameter hole and using water with a pressure of 55 psi, I estimate a flow rate of 0.86 gpm.

This calculated flow rate could be easily confirmed by drilling a single $\frac{1}{16}$-inch-diameter hole in a piece of pipe, turning the water on, and timing how long it takes to fill a 1-gallon jug.

Assuming my calculation is close to correct, and that you have a water source that will provide up to 5 gpm, you could drill only five small holes before you start dropping the pressure by creating more openings than the source is able to supply water at full pressure. Because a $\frac{1}{32}$-inch hole is only a quarter of the area of a $\frac{1}{16}$-inch hole, you'd probably be able to drill four times as many $\frac{1}{32}$-inch holes and still

maintain pressure if you could make your holes that small.

Maintaining the pressure is important for this application as it's both the speed and the quantity of water hitting the roots that makes a difference in how effectively they're cleaned. Keep in mind that once the water exits the hole it hits the air, which starts to slow it down. This is why getting the nozzle close to the roots is important when trying to get them clean with as little water as possible.

Here's a more in-depth explanation of how I calculated that flow rate:

$$v = \sqrt{2 \times head \times g}$$

To find the speed of water exiting a hole, you can use the above formula (known as the Bernoulli equation), where v is the speed (or velocity), g is the acceleration due to gravity (32.17 feet per second^2), and the head elevation is the pressure. (See Fluid Dynamics on page 196 for an explanation of converting psi to head.) For a pressure of 55 psi, the exit speed of water will be a little over 90 feet per second.

If the diameter of the hole is $\frac{1}{16}$ inch, the area of the hole is $\pi \times r^2$, or 0.0031 inches^2. To use this number with the speed, convert it to square feet (0.000021 feet^2).

Multiplying this area by the exit speed gives you a volume per time: 0.0019 feet^3 per second. You now need to convert the cubic feet to gallons and the seconds to minutes to end up with 0.86 gpm (depending on the number of decimal places you carry for those calculations you might end up with 0.87 gpm).

USING THE MINI BARREL WASHER

To use the Mini Barrel Washer, set it on a pair of sawhorses or some other support and make the entry end a bit higher than the exit. Screw a garden hose to the pipe, make sure the gate is closed, and dump a load of roots into the entry ramp. Turn the water on, then rotate the barrel by hand (or you can motorize the barrel; read on). Rotating the barrel will tumble the roots toward the gate. You can adjust how fast the roots tumble by adjusting the height of the support under the entry end; a steeper angle will make them move faster. Once the roots are clean, position a clean container under the exit ramp, open the gate, and let the roots tumble out.

Even though this was intended to be a prototype, ultimately it worked well enough that Danny is still using it, with a few modifications, seven years later. For a few years Danny rotated the barrel by hand. It didn't actually save much time, but it did get the roots nice and clean. Eventually he added a general-purpose ¼-horsepower DC motor with a built-in 33:1 gear reduction and a controller that allows him to vary the speed.

The electric motor is mounted on a wooden frame built above the middle of the barrel. A cog driving a roller chain that loops around the outside of a third rim, which was added to the middle of the barrel, transfers the torque from the motor and rotates the barrel. Having a motor means Danny and his crew can put in a load of roots, turn on the water and the motor, and let the load tumble for a few minutes while they do something else.

My prototype barrel wasn't super strong because I initially relied only on the existing spoke holes in the bicycle rims, and some of the screws holding the boards to the hoops came out. In addition, some screws stuck through the boards and damaged roots. In the years that

Danny has been using the barrel washer, he has replaced a lot of the screws and the caster wheels once, after they began showing some wear.

To improve the barrel washer, you could divert the water coming off the bottom of the barrel, away from the wash area. I've seen many farms divert the water with corrugated roofing bent into a slight arc positioned under the barrel. In addition, if you divert the water into a big tub used as a settling tank, you could let the majority of the soil settle and then recirculate the water to the barrel entrance. This way you could give the roots an initial wash with recirculated water before a final rinse with clean water, potentially saving water and improving the speed of the washer.

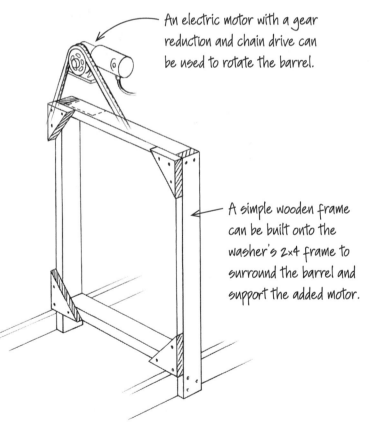

An electric motor with a gear reduction and chain drive can be used to rotate the barrel.

A simple wooden frame can be built onto the washer's 2x4 frame to surround the barrel and support the added motor.

DESIGN NOTES

While the basic idea for the washer is pretty simple, I did put some thought into optimizing the support structures and minimizing materials use. I also used a few basic calculations. For example, the interior diameter of the rims was about 21¼ inches. If you remember your geometry, circumference = diameter × π (3.14). Therefore, 21.25 × 3.14 = 66.75. That calculation let me know that the inside circumference of the wheel rims was 66¾ inches.

The cedar fence boards were about ⅝ inch thick and 5½ inches wide, so I needed about eleven segments to fill the circumference, leaving a total gap of about 5¼ inches. To distribute that gap evenly, I left about ½ inch between each board. Because the boards are straight across and didn't actually follow the curve of the rim, and they're ⅝ inch thick, the gap between boards ended up being slightly less than ½ inch. This space is large enough to let water and soil through, but small enough to prevent roots from falling through.

The rims were already drilled for spokes, so I screwed through the existing holes to secure the boards along the inside of each rim. Attaching the boards in this way results in an all-wood barrel interior, with no metal contacting the crop. The connection isn't super strong, but it's enough to hold the boards in place. In normal use, most of the force on the boards pushes out toward the hoops, so there isn't much strain on the screws. But if you push on a wet board from the outside in, you can easily break the connection.

The predrilled holes in the rims line up with the boards inconsistently, though. I recommend the additional step of drilling extra holes in the rims so that you can consistently set two screws per board, with each screw near the edge of a board.

You could rip the boards into narrower strips that would more closely follow the curve of the rim and match up with the existing hole spacing, but there is some advantage to the more angular interior you get from using wider boards. The angles cause the roots to tumble a little more, helping get them cleaner. On a smoother surface they just tend to slide rather than exposing new parts of the root to the water and rubbing up against each other.

Two additional factors worth considering are the length of the barrel and the placement of the rims. The longer the barrel, the more time the roots spend inside getting washed off as they tumble from one end to the other. A longer barrel also means more room for spray heads. And while additional spray heads mean more water—which does result in more cleaning—Danny's shed had a limited amount of water flow and space, so I chose to keep the barrel shorter. A shorter barrel also makes it easy to reach in from either end to clear debris or clean the barrel without having to crawl in or use a long-handled tool. My arms are about 2 feet long, so a 4-foot barrel seemed reasonable.

Not only do the bicycle rims serve as the support hoops that hold the barrel together, they are also the tracks that the caster wheels hold up. In general, spacing the hoops farther apart stabilizes the barrel. If you position the hoops too far apart, though, the middle section of the barrel won't be supported. I set each hoop about 9 inches from the barrel edges (about one-fifth the total board length). This spacing gives a little extra support to the ends, where the boards are supported by only one hoop, and a little less support to the center of the barrel, but the center gets some support from both hoops.

The entry ramp on this barrel washer is designed for the roots to be poured in from a harvest tote without being bruised or lost and then to softly exit the barrel to a clean tote. I made the entry ramp load from the side. This simplifies the transition from the flat lip on which the harvest tote rests to the round funnel that feeds the barrel. Because this section is slightly tilted, the roots tumble into the rotating barrel with only a little push. This makes it easier to control how fast the roots are fed into the barrel.

On the exit end I put a flat ramp just below the rotating barrel and added a gate that can be slid open and closed. Danny removed this ramp and now the clean vegetables empty straight into a tote. You could also have the barrel empty onto a sorting table, which is very common.

The closed gate keeps the roots tumbling in the barrel, giving them more time to be washed off. You can adjust the amount of time the roots take to tumble from one end of the barrel to the other by changing the angle of the barrel. Simply adjust the height of one end of the frame: a steeper angle will move the roots through the barrel more quickly, and a shallower angle will move the roots more slowly (or not at all). While I haven't designed legs for this barrel washer, the base frame could have legs added to it or sit up on top of sawhorses.

The last part of the washer design is the water spray system. The garden hose in Danny's shed provided about 5 gallons of water per minute, so in order to maintain good water pressure it needed spray nozzles that put out less than

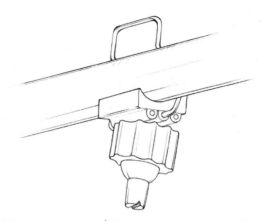

Clip-on nozzles aren't the easiest to source, but if you can find them they are simple to install and to remove for cleaning or replacement.

5 gallons per minute when all used together. To make the direction and angle of the nozzles adjustable, I purchased K-Ball nozzles that mount by drilling a hole in a pipe and clipping the nozzle around the pipe. Once installed, these nozzles can be rotated to direct the spray. This mounting approach saves money and time and minimizes disturbances to flow along the length of the pipe because extra fittings, like tees, are not needed.

Each of the nozzles has a flow rate of 1½ gallons per minute, so I was able to space out three along the length of pipe. If you install too many nozzles and exceed the maximum flow rate available, the pressure will drop. When pressure drops, water exits the nozzles more slowly. Water (or anything) moving at a higher speed has more energy, and when water with more energy hits an object (like dirt on a root) it exerts more force on that object than water moving more slowly.

ROLLING PACKING TABLE

In the packing shed, we always need a clean surface that is at a comfortable height for sorting, packing, and labeling boxes of produce. Packing and labeling involves lots of supplies, like permanent markers, masking tape or labels, clipboards full of orders and packing instructions, wax boxes, and other packing materials; all of these need a place to live and need to stay dry.

This custom table helps with all of it: The tabletop is adjustable so it can be set at the right height for specific workers or tasks; a narrow shelf above the tabletop stores small supplies and keeps them dry; a larger shelf under the tabletop provides storage for wax boxes and other large packing supplies; and the table is on casters so it can travel wherever it's needed.

In addition to moving the table to where it is most needed, the wheels allow you to use the table for transporting materials and product around the shed. And two of the four caster wheels can be locked, preventing the table from moving when you want it to stay in place.

Most of the table is made with standard lumber and plywood. The table surface is a prefab plastic folding table (see Design Notes on page 146 for sources for the table and the casters). If you want to use a different type or size of table surface, you can simply custom-fit the wood structure around the tabletop.

I rated the level of complexity for this project as high not because it's particularly complicated but because it requires some attention to detail to ensure that everything fits together well. Because the tabletop needs to slide for height adjustment, but also shouldn't be so loose that it wiggles excessively, take care to get the cuts and attachments just right.

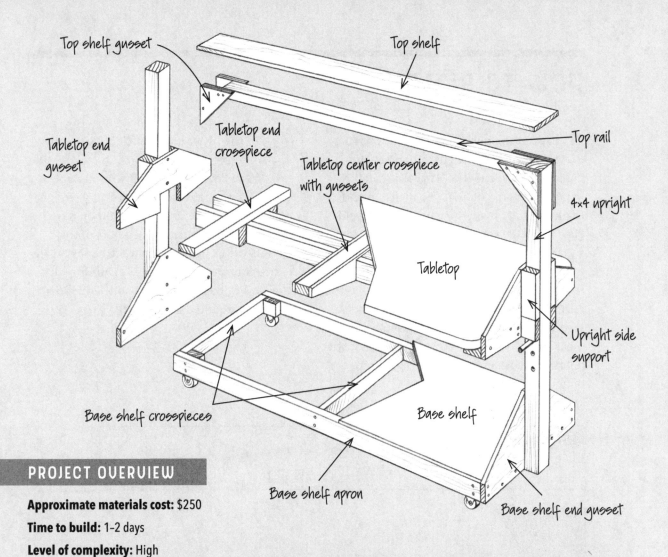

Top shelf gusset

Top shelf

Tabletop end crosspiece

Tabletop end gusset

Top rail

Tabletop center crosspiece with gussets

4×4 upright

Tabletop

Upright side support

Base shelf crosspieces

Base shelf

Base shelf apron

Base shelf end gusset

PROJECT OVERVIEW

Approximate materials cost: $250

Time to build: 1–2 days

Level of complexity: High

SUGGESTED TOOLS

- 48" straightedge
- Framing square
- Circular saw or table saw
- Jigsaw
- Drill
- Trigger clamps
- Hacksaw or grinder with abrasive wheel
- File or grinder
- Center punch
- Paperboard for shims (see step 14 on page 147)

RECOMMENDED MATERIALS

One 48" × 96" sheet of ½" plywood

Eight 96" lengths of 2×4s

One 96" length of 4×4

Two 3" double-locking casters

Two 3" fixed casters

One ½" × 12" black pipe nipple

Two ½" black pipe caps

Two 3⁄16" lynchpins

1½", 2", and 2½" deck screws

Sixteen ¼" × 2" panhead wood screws

80-grit sandpaper

6' plastic folding table or equivalent (see Design Notes on page 146)

NOTE: Be sure to select a 4×4 that is straight and has no twist. The table slides on the 4×4 uprights, and it will bind if they are not straight and true.

HOW TO BUILD IT

1 Lay out the plywood parts. Lay out all of the cuts for the gussets and shelf pieces on a full sheet of ½" plywood, following the diagram below. If you measure and mark carefully, everything will fit on the single sheet. Use a straightedge and a framing square to draw accurate lines.

NOTE: The actual dimensions of the tabletop I used were 30" wide and 71½" long, and the step-by-step instructions are based on those measurements. If your table is a different size, follow the cues in the steps to customize the dimensions for your table.

2 Cut the plywood parts. Cut out the plywood pieces, using a circular saw or table saw (if you have one) for the long straight cuts. Be sure that these cuts are very straight and smooth (see Cutting Plywood on page 113 for tips). Cut the inside corners of the gussets with a jigsaw. Soften the edges on all of the cuts and remove any splintering with a few quick passes using 80-grit sandpaper wrapped around a wood block.

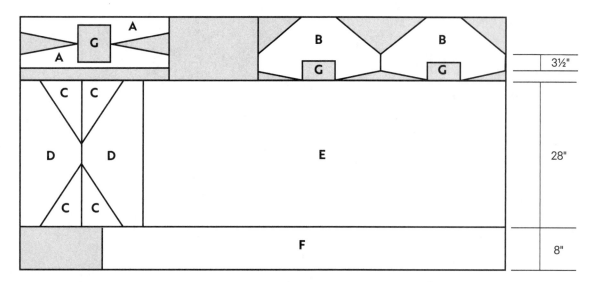

Diagram parts: (A) tabletop center gussets; (B) tabletop end gussets; (C) top shelf gussets; (D) base shelf end gussets; (E) base shelf; (F) top shelf; and (G) cutouts in tabletop gussets. All of the shaded areas, including the three pieces labeled G, represent unused materials.

When laying out the plywood pieces, the most critical dimension is the length of the base shelf (E). It should be a hair longer—up to ⅛"—than the tabletop so that the tabletop can slide easily between the uprights. The top shelf (F) is the length of the tabletop plus 8", to account for the width of the two 4×4 uprights and the top

shelf gussets (C). The rectangular cutouts (G) in the tabletop gussets (A and B) should be 3½" × 6½" to fit over the two 2×4 rails sandwiching the 4×4s.

All of the angles on the gussets should approximately line up with the edges of the pieces to which they will connect. The precise angles are not critical.

3 **Cut the 2× pieces.** Cut the 2×4s as follows:

- Two 71½" base shelf apron pieces (these pieces should be a hair longer—up to ⅛"—than the length of your base shelf)

- Three 25" base shelf crosspieces

- Two 79½" tabletop rails

- Four 8½" upright side supports

- Two 24½" tabletop end crosspieces (these pieces should be just a little shorter than the width of the table at its very ends; my table has rounded corners, so I measured only the straight edge at the ends)

- One 29½" tabletop center crosspiece (this piece should be just a little shorter than the actual width of your table at its center)

- Ten 3½" blocks

4 **Assemble the base shelf frame.** Lay out the base shelf apron pieces and crosspieces to make a frame, following the illustration on page 141. Drill two pilot holes per connection in the apron pieces to prevent splitting the wood (see Driving Screws on page 11). Screw the bottom frame pieces together using 2½" screws; be sure to get the end crosspieces exactly flush with the ends of the apron pieces. Make sure the frame is square at the corners, then attach the plywood with 1¼" screws every 12"–16" around the edges and through the crosspiece in the middle of the frame.

TIP ALIGNMENT

The base shelf crosspieces should be flush with or very slightly proud of the plywood shelf to allow the uprights to sit vertical without splaying out at all. Also, if the base shelf platform assembly is shorter than the tabletop, the sliding mechanism won't work, so double check that it is at least as long as the tabletop when assembling the frame.

5 **Attach the wheels.** Set one of the 3½" blocks in each inside corner of the base shelf frame, flush with bottom edge, and secure each block with a single 2½" screw. Use one set of casters to mark the hole locations for the screws, then drill pilot holes for ¼" panhead screws. Fasten the two double-locking casters onto one end of the base and the fixed casters onto the other end.

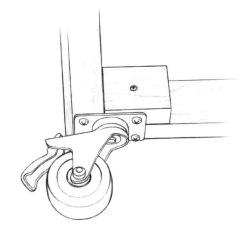

6 Attach the base shelf end gussets. Center the end gussets along the base shelf crosspieces using two 1½" screws at each outside corner and four more screws along the length, avoiding the area where the 4×4 uprights will go.

7 Cut the 4×4 uprights. Cut the 4×4 in half to create two equal-length pieces. The actual length is less important than making sure they are equal in length and that the cuts are square, especially on the top ends. With most circular saws this will require two careful cuts from opposite sides to make it through the full thickness.

8 Drill the holes for the pins. Drill holes in the 4×4s for the ½" black pipe pins that will hold up the tabletop. If you have a drill press that will make a hole 3½" deep, use it to help keep the holes square. To determine where to place the centers of the holes, you need to decide what table heights you'll want to use and then do a little math. See the tip below or simply make a range of heights for the table starting by placing a hole 12" above the bottom of the 4×4 and drilling another hole every 2" or 3" up to 30", depending on how much adjustment you want. Use the holes in one 4×4 as a template for marking the other 4×4 so that the hole locations will be exactly the same and the tabletop will sit level.

TIP DRILLING HOLES IN THE 4×4S

When drilling the holes in the 4×4 uprights, I used a ⅞-inch spade bit on a hand drill and was very careful to keep it square to the workpiece. If you use a spade bit, drill most of the way through the 4×4, but stop just when the point of the bit comes through the other side of the wood. Then drill from that side, using the small hole to center the bit, to complete the large hole. This prevents the bit from making a messy, splintered hole at the far side of the wood, and it keeps the bit aligned so that the two holes from opposite sides line up properly.

To decide where to place the holes, you need to decide on the table heights that will be most useful for those who will work at the table. I generally like to have the tops of boxes that I'm sorting or packing a little below my elbow height. We have two distinct box heights, and sometimes we also pack boxes that are sitting on a scale. If more than one person will be using the table, take measurements for the tallest and shortest person to get a likely range.

An example of the measurement calculation would be using an elbow height of 41 inches and subtracting 2 inches (to get a bit below the elbow), the box height of 9 inches, and the scale height of 6 inches. This gives a target table height of 24 inches. To make a hole that gives you that table height, subtract 9 inches from 24 inches to compensate for the casters and the tabletop, rails, and crosspieces. This gives you 15 inches. The hole then gets drilled 15 inches up from the bottom of the 4×4. If I want to use the same box without the scale I'd drill a second hole 6 inches higher so that the top of the box remains at the same height when I'm working with it.

9 **Cut the top shelf rail.** Cut the 2×4 top shelf rail to exactly match—or come just shy of—the length of the assembled bottom frame, including the two plywood end gussets. If the top shelf rail is too long, it will be impossible to make the uprights square.

10 **Attach the 4×4 uprights to the base shelf end gussets.** Center the 4×4s on the gussets and make sure they are square to the base shelf platform. Drive two $1\frac{1}{2}$" screws through each gusset and into the 4×4, one close to the top of the gusset and the other low, close to the platform. Wait to put in more screws until the joints at the top of the 4×4s are secured in the next step.

11 **Secure the top shelf rail.** Fasten the top shelf rail in place, flush with the tops of the 4×4 uprights, with the triangle gussets on either side, using clamps to hold the work as you go. Check that every joint is square, then tack the parts together by driving two $1\frac{1}{2}$" screws through each gusset, one into the rail and the other into the upright. Double-check to make sure everything is square, and then stiffen the frame by adding four or five screws on each edge of each triangle. Then stiffen the rest of the frame by adding two to four screws through the base frame gussets and into the 4×4s, as well as two to four $2\frac{1}{2}$" screws through the 2×4 base shelf crosspieces, through the gussets, and into the 4×4s.

12 **Add the top shelf.** Center the plywood for the top shelf along the top rail and attach it, starting with two $1\frac{1}{2}$" screws on each end at the corners of the 4×4. Then zigzag screws down the length of the shelf, through the plywood and into the top rail every 10"–12".

TIP SQUARING THE UPRIGHTS

I find that the easiest way to square an upright, especially if you're building the table solo, is to get the upright close to square and then clamp it in place with a trigger clamp. Drive one screw, then slightly loosen the clamp, adjust the upright as needed to square it perfectly, reclamp, and drive the second screw.

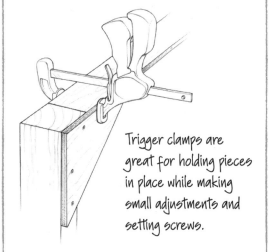

Trigger clamps are great for holding pieces in place while making small adjustments and setting screws.

DESIGN NOTES

The 6-foot plastic folding table that I used for a table surface is durable and easy to clean. The model I used is the 30 × 72-inch heavy-duty folding table from National Public Seating, but similar ones are available from many different manufacturers. Alternatively, the plastic table could easily be exchanged for a stainless kitchen work surface, which would add a couple hundred dollars to the cost of materials (for a new surface), or a plywood top, which would be very functional but might be harder to clean.

Other than being screwed to the frame so that it doesn't come loose during use, the plastic folding table is not modified. It can be removed and used as a regular table again if needed, or replaced if the surface wears unacceptably.

Finding the right balance so that the tabletop assembly slides easily but doesn't wobble can be delicate. I haven't used this design long enough to know how much the wood's natural expansion and contraction—with variations in humidity and temperature—affect how easily the tabletop slides. Constructing the tabletop at a fixed height would greatly simplify the construction, but it would reduce the potential benefits to workers of different heights. A fixed tabletop would also make it less convenient to use this piece for multiple tasks that call for different working heights.

I chose to use two fixed wheels and two double-locking casters for a few reasons. Double-locking casters don't roll or pivot when locked, but they do wiggle quite a bit. Using four double-locking casters would make it easier to roll the cart into position in tight spaces, but you would then need to lock four casters instead of two. Even locking only one caster in this configuration keeps this cart pretty stable.

The casters I used are from Peachtree Woodworking Supply, but there are many suppliers that carry similar ones. Pay attention to the load capacity and wheel size when selecting casters. Wheels with a higher load capacity tend to last longer and make the table safer because any single wheel could carry the full weight of the table and everything you might ever put on it. Larger wheels roll more smoothly but take up more height.

13 **Make the pins.** Cut the black pipe nipple in half with a hacksaw or a grinder with an abrasive wheel. Soften the edges of the cut with a file or a grinder. To drill holes for the lynchpins, use a center punch to mark a hole about ½" in from the cut end of each pipe; this will keep the drill bit from wandering as you start to drill. Use a $\frac{1}{8}$" bit to drill a pilot hole (see Drilling Metal on page 16). Widen the pilot hole with a ¼" bit, using a handheld drill with the pipe clamped firmly to the table, or use a drill press. Finish assembling the pins by screwing the caps onto the threaded ends.

14 **Build the tabletop support platform.** This step requires a bit of fiddling in order to find a good balance, one that lets the tabletop slide easily but without wobbling excessively.

- Insert the pins into the top holes in the 4×4 uprights and rest the long tabletop rails on top of the pins, holding them loosely to the 4×4s with trigger clamps so they don't fall off.

- Slide a thick piece of paperboard or, even better, three thin pieces of paperboard (from a cereal box, for example) between one of the rails and the 4×4s to shim it out slightly, making sure the shim sticks out a bit so you can easily remove it when you're finished. (If you have three pieces, the middle piece will be the easiest to pull out first, then the other two should come out easily. Don't let the shim material stick up into the area where the tabletop end gussets will go.)

- On each end, fit the tabletop end gusset over the two 2×4 rails, then clamp them loosely to the 4×4 uprights to keep them vertical. Use a paperboard shim behind one of the two gussets to set it slightly off the 4×4.

The black pipe nipple has a threaded pipe cap on one end and a lynchpin on the other end.

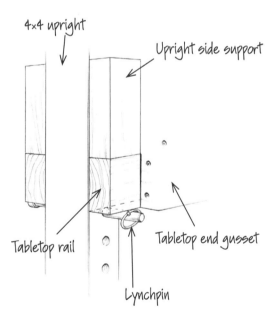

4×4 upright

Upright side support

Tabletop rail

Lynchpin

Tabletop end gusset

The metal pin sticking through the 4×4 upright holds the sliding tabletop assembly at the desired height. The lynchpin keeps the larger metal pin from accidentally coming out and is easy to remove to allow quick adjustments to the table height.

- On each end, clamp one 3½" block to the outside of each rail, flush against the inside of the gusset, with the grain running vertically, as shown below. Predrill two holes through the side gussets and into the blocks, then screw the blocks in place with 1½" screws. Predrill a single hole through the center of each block into the rail and secure it with a 2½" screw. Then screw the last pair of blocks to the center of each rail, on its outside face, again with the grain running vertically. (You'll fasten the center crosspiece and gussets to this last pair of blocks in the next step.)

- Center the three 2×4 crosspieces across both rails, with the longer piece in the middle. Screw the end crosspieces to the end gussets using three or four 1½" screws per side through the gussets. Add the center gussets to the center crosspiece using the same screw attachment method you used with the end gussets. You should end up with 10–12 screws going through each gusset.

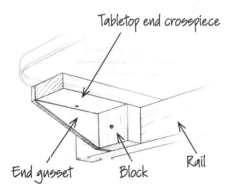

A 3½" block is set flush up against the end gusset and screwed to the rail under the tabletop. To create a stronger joint and help prevent splitting, be sure that the grain in the block runs vertically and that all screw holes are predrilled.

- Set the 8½" upright side supports flush against the end gussets, using paperboard shims between the support pieces on one side of the table and the 4×4 uprights. Clamp to hold everything in place. Tack the upright side supports in place by driving a single screw through each gusset into the top of each side support. Then toe-screw at an angle through the side supports into the rail on the opposite side of the gusset to keep the side supports from twisting. Add one or two more screws through the gussets into each side support to lock it all together tightly. At this point you can remove all of the paperboard shims. The tabletop support assembly should slide up and down on the 4×4 uprights.

15 **Fasten the tabletop.** If you've measured everything correctly, the table with its legs folded flat should slide onto the crosspieces snugly, or at least come close to the end gussets. Center the table on the crosspieces, then drive 2" screws (or other screws that are just long enough by a few threads to penetrate the plastic but short enough to not stick out the top of the table) through the bottom of the crosspieces up into the bottom edge of the table. The screw tips will prevent the table from sliding off the support assembly. These screws can be removed easily if you want to use the tabletop as a regular table or replace it when the surface is excessively worn.

USING THE ROLLING PACKING TABLE

To set the table height simply put the pins in the proper holes. This should be done when there is nothing on the tabletop. It is probably easiest to adjust the tabletop height with one person on either side, but you may be able to adjust one side at a time without trouble.

There are many ways you can customize this table to accommodate special needs, such as adding places to hang clipboards or hold tools, adding special shelves or outlets for electronics or scales, or creating dividers below for different box and bag sizes or other packing materials.

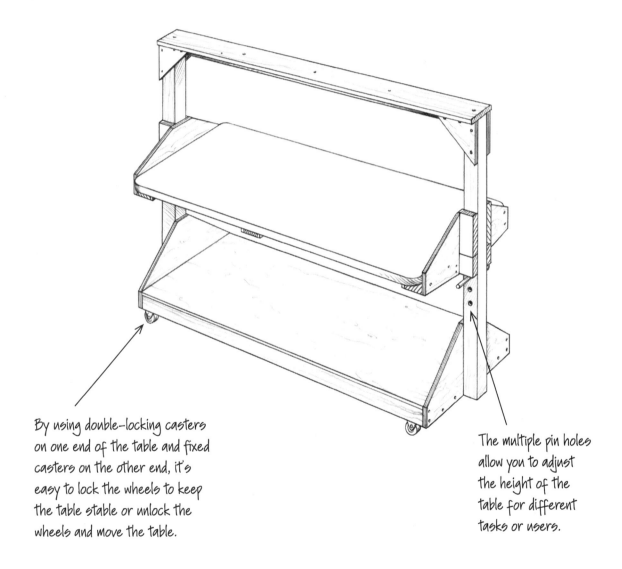

By using double-locking casters on one end of the table and fixed casters on the other end, it's easy to lock the wheels to keep the table stable or unlock the wheels and move the table.

The multiple pin holes allow you to adjust the height of the table for different tasks or users.

CSA BOXES

I used these practical wooden boxes to deliver bags of farm produce to our CSA members when I worked at Our Table Cooperative in Sherwood, Oregon. The individual CSA shares were packed in canvas tote bags, but I needed a way to transport the bags to the pickup sites and to keep them protected until people picked them up.

These boxes are fairly lightweight and, importantly, they are easy to carry. When stacked, each box locks on to the lid of the box below, which makes the stacks more stable for transport. A lot of folks use plastic totes, but I wanted to get away from the look and feel of plastic, especially for what was largely a display box.

This is one of the trickiest tools in the book to build, mostly because of the level of accuracy required. For that reason, you will want to use a table saw for cutting the plywood accurately. A router is helpful for milling slots in several of the pieces and for softening edges. It also helps to have a miter saw, a drill press with a 4-inch hole saw, and a pneumatic stapler to speed up the process if you're making larger batches of boxes; these aren't essential (you can improvise with other tools, if necessary), but they'll help you produce a better box with less effort.

TIP A NOTE ABOUT MATERIALS

Use quality lumber and plywood for these boxes. If you can find a five-ply ¼-inch plywood with good surface veneer, the boxes will be easier to put together and will wear much longer. I've tried making the boxes with cheaper materials and they did not hold up well. You might look for Baltic birch plywood, which is very high quality but typically comes in nonstandard sheet sizes, so you'll have to adjust for that. It's also worth noting that ¼-inch plywood varies in its actual thickness and will usually be closer to $\frac{3}{16}$ inch thick. This design takes that into account.

If you're building more than one batch of boxes, save the leftover materials from the first batch; you should be left with half of one 1×6 and most of a ¼-inch and a ¾-inch piece of molding.

The molding is sometimes referred to as stop molding, and it should have a rectangular profile with at least one broad, flat side that is the contact side. The other side usually has slightly dulled corners, which actually work well for this application. Avoid using a stop molding with a fully rounded corner or any kind of more elaborate profile.

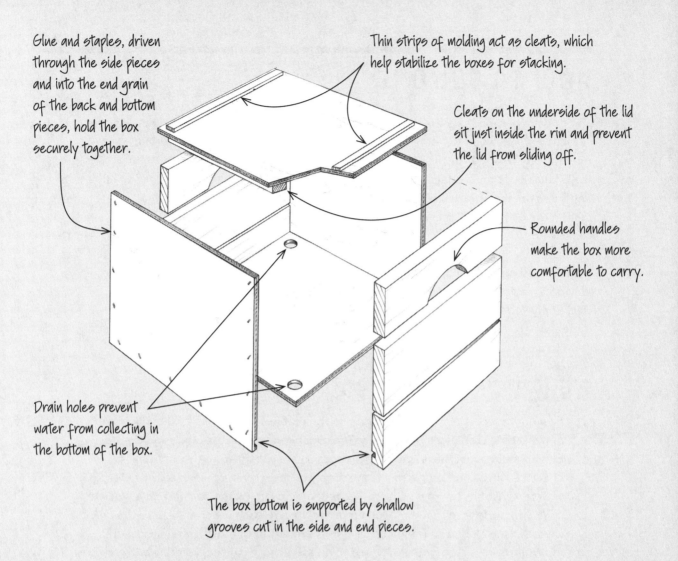

Glue and staples, driven through the side pieces and into the end grain of the back and bottom pieces, hold the box securely together.

Thin strips of molding act as cleats, which help stabilize the boxes for stacking.

Cleats on the underside of the lid sit just inside the rim and prevent the lid from sliding off.

Rounded handles make the box more comfortable to carry.

Drain holes prevent water from collecting in the bottom of the box.

The box bottom is supported by shallow grooves cut in the side and end pieces.

PROJECT OVERVIEW

Approximate materials cost: $12-$15 per box

Time to build: 2-4 hours for one, or 30-45 minutes per box in larger batches

Level of complexity: High

SUGGESTED TOOLS

- Table saw
- Miter saw (optional but helpful)
- Router (with a router table, if you have one) with $1/4$" roundover bit and $3/16$" straight bit (optional)
- Drill (preferably a drill press) with a 4" hole saw and $3/8$" bit
- Stapler (preferably pneumatic)

RECOMMENDED MATERIALS
(for a batch of 9 boxes)

Two 48" × 96" sheets of $1/4$" plywood
Five 96" lengths of 1×4
Three 96" lengths of 1×6
Three 96" lengths of $1/4$" × 1" molding
Three 96" lengths of $3/4$" × 1" molding
80-, 100-, or 120-grit sandpaper
$1/4$" × $3/4$" crown staples
Wood glue

HOW TO BUILD IT

1 **Cut the plywood pieces.** For each box, cut from plywood two sides, 13" × 15¾"; one bottom, 12⁹⁄₁₆" × 14⁷⁄₁₆"; and one top, 12¾" × 15¾". Take your time setting up your table saw and be as precise as possible with all of your cuts. In the end this will make the boxes go together much more easily.

NOTE: These measurements assume your plywood is actually ³⁄₁₆" thick. If you get slightly thicker plywood, you may need to adjust the dimensions to fit.

2 **Cut the 1×4 pieces.** For each box, cut four 12⅜" lengths of 1×4.

3 **Cut the 1×6 pieces.** For each box, cut two 12⅜" lengths of 1×6.

4 **Cut the molding.** For each box, cut two 12⅜" strips of the ¼" × 1" molding and two 14³⁄₁₆" strips of the ¾" × 1" molding.

TIP USING A STOP BLOCK

If you have access to a miter saw or a sled for a table saw, set up a stop block instead of measuring each piece individually. The distance between the blade and the stop block should equal the length of the piece you are cutting from the molding and the 1× lumber. If you are using a miter saw, simply clamp a scrap of wood to the fence, or if your fence isn't as long as the piece you're trying to cut, fix the saw and stop block to your work surface so that they won't move relative to each other. Make sure the block is square to the blade and is tight enough to prevent it from moving while you work.

When using a stop, you don't need to measure any of the cuts; simply butt the lumber up against the stop block, hold it tightly, and make the cut. This ensures that all of the cut pieces are exactly the same length.

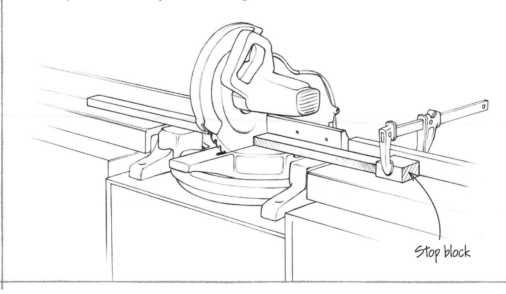

Stop block

5 **Cut the handles.** To make comfortable handles, use a 4" hole saw to cut out little arcs from one long edge of each 1×4. You can cut two pieces at a time, using a simple jig made with a scrap of plywood and a few blocks of wood. The jig holds a pair of 1×4s in place while the hole saw cuts both 1×4s at the same time. Use a spacer block cut from a scrap of wood to hold the two pieces approximately ¾" apart so that the hole saw will cut a bit less than a half circle in each piece.

6 **Sand the cut edges.** Quickly sand off any splinters raised by the saw blade on the ends of each 1× piece and the plywood. Smoothing the ends keeps the splinters from getting worse and from getting in your way later. For this step, I simply wrap a piece of sandpaper around a scrap block of wood and take three or four strokes, sanding off the end of the board (as opposed to sanding toward the board, which may create more splinters).

7 **Cut slots in the side and end pieces.** Use a router table with a ³⁄₁₆" bit to cut a ³⁄₁₆" slot in both plywood side pieces and in the two 1×4s without handle cutouts. Position these slots ¼" from the bottom edge (one of the long edges) on each piece, on the inside face (most plywood has a better-quality ply on one face, and you'll probably want that side facing out). The slots should be ³⁄₃₂" deep (about half the thickness of the plywood), so take some time to set the bit exactly right. Accuracy here will help ensure that each piece of the box fits together securely for the assembly.

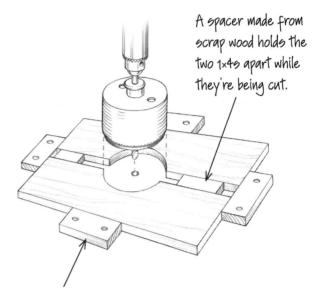

A spacer made from scrap wood holds the two 1×4s apart while they're being cut.

Screw four blocks to a scrap of heavier plywood to make a nice jig. Secure the jig to the hole saw's table with two clamps or bolts.

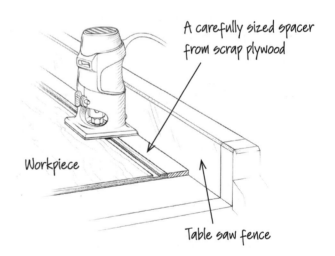

A carefully sized spacer from scrap plywood

Workpiece

Table saw fence

If you don't have a router table, you can use your table saw fence with a spacer carefully cut from the same plywood to get the router to cut a groove in the correct place. For additional tips on cutting slots, see the next page.

8 **Drill drain holes.** Drill a ⅜" drain hole in each bottom plywood piece approximately ½" in from each corner. These holes keep water from collecting in the bottom of the box when you wash it out or put in wet produce.

TIP CUTTING THE SLOT

If you don't have a router table, here are a couple of other good ways to cut the slot.

Use the fence on your table saw as a straightedge for a handheld router. Cut a strip from a scrap piece of the same plywood you are using for your boxes. Make this strip exactly wide enough that when it sits between the fence and your workpiece, your router bit will be positioned ¼ inch from the edge of the workpiece. Hold the workpiece against the spacer that is up against the fence, then run the router along the fence to get a straight slot.

Make two passes with a table saw. Standard table saw blades are ⅛" wide. To get a wide-enough groove for your slot, set the fence so the blade cuts one groove edge closest to the bottom edge and make all of those cuts. Then reset the fence to cut the other side of all of the grooves, overlapping your passes slightly. This doubles the time that it takes to cut the grooves, but if you don't have a router and are making only one batch of CSA boxes, this approach is almost certainly less expensive than going out and buying new tools.

9 **Assemble the box bottom.** For each box, gather the plywood bottom piece and the four pieces that you just cut a slot into (two plywood side pieces and two 1×4s). You're going to slip the bottom piece into the grooves in the other pieces and secure it with glue. The assembly is somewhat involved, so first practice it without glue or staples to confirm that the pieces fit together and that you understand the process. Once you start applying glue, you'll have to work quickly to assemble each box bottom so that the glue doesn't dry out.

When you're ready to assemble with glue, apply a bead of glue to the slots of the 1×4s and to the slot of one of the side panels. Slip the 1×4s over the bottom panel and stand this all up on the remaining side panel (the one without glue in the slot) to hold the pieces upright. Add a bead of glue to the end grain of each 1×4 piece, and place the side panel with the glued slot on top.

Holding all of this together tightly, drive three evenly spaced staples through the side panel and into the end grain of each 1×4. Flip over the pieces that are stapled together and put a bead of glue in the slot of the unglued side panel and along the end grain of the 1×4s. Press the side into place and staple it, as with the other side panel. If you're making a batch of boxes, assemble all of the bottoms before going on to step 10.

NOTE: Use enough glue that it makes contact with both mating pieces, but not so much glue that it squeezes out when you join the pieces. Just to be safe, keep a damp rag on hand to wipe off any excess glue that squeezes out during assembly.

10 **Finish assembling the ends.** Place a thin bead of glue on the end grain of the other 1× pieces and slide them into place as follows: The 1×6s go in the middle and the handle pieces go on top, with the handle cutouts facing downward, toward the 1×6s. The tops of the 1×4 handle pieces should sit flush with the tops of the side panels. Assemble one end at a time, leaving a slight gap between each 1× for ventilation. When both of the pieces are in place, add three staples per end to both 1×s. Repeat with the other end of the box.

11 **Ease the edges.** If you have a router with a $\frac{3}{16}$" or $\frac{1}{4}$" roundover bit, run it over the inside and outside edges of the handle cutouts and along the inside perimeter of the top edges of the box. If you don't have a router, use sandpaper to soften the edges. Both approaches will make the handles more comfortable to grip and the lids easier to put on.

12 **Assemble the lids.** The lids are actually quite tricky to assemble even though they look relatively simple. The tricky part is lining up the molding pieces that will act as cleats. Getting these just right is important for a good fit, and you need to be able to do both sides of the lid at the same time. If you're just making one box you can probably fiddle around and get it right without a jig. If you're making more than one or two, however, I highly recommend building yourself a jig with spacers that will hold the bottom cleats ($\frac{3}{4}$" × 1" molding strips) in place, locate the top panel, and then hold the top cleats ($\frac{1}{4}$" × 1" molding strips) in place. (See Constructing an Optional Lid Jig on the next page for information about the jig.)

Use a bead of glue to attach the two longer, thicker pieces of molding to the bottom of the lid exactly one width of your plywood away from the long edge of the lid and centered along that length. (This keeps the lid tightly in place when it's on top of the box.) Place the two shorter pieces of molding on top of the box exactly the width of a 1× away from the short edges of the lid and centered along that length. (These prevent a box that is stacked on top from sliding sideways.) Add a bead of glue to each of these cleats. At each corner, where the cleats overlap, place one staple that goes through the top cleats and the lid and down into the bottom cleats. This will hold them in place while the glue dries.

CONSTRUCTING AN OPTIONAL LID JIG

To make the bottom part of the jig, cut two 1×4 pieces the same length as the ones used in the boxes and two strips of the same plywood you used for the box sides measuring slightly more than 4¼" wide and about 10" long. The exact dimensions for the plywood aren't important, but having the two long edges parallel and one end square to the long edges is important.

Staple one of the 1×4 pieces to the square end of the plywood strips, mimicking the top end of a box. Position the second 1×4 at the other end of the strips, sitting the same distance below the edge as your ¾" molding is thick. To get this spacing right, flip the partially assembled jig over on a flat work surface and place two pieces of the ¾" molding for the box lids along the inside edges of the plywood, broad edges down. These will be your spacers. Place the second 1×4 on top of the molding and staple through the plywood and into the 1×4. Use two staples per side to hold it in place.

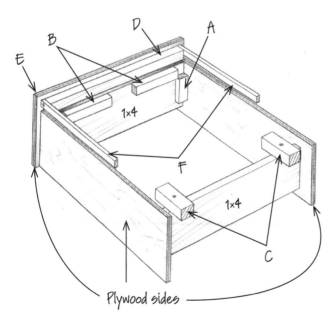

You can make the lid jig out of scraps from the same materials you used for the boxes in order to match the necessary dimensions easily.

While you have the whole thing upside down, slide the two ¾" molding spacers into the corners so that they butt up against the 1×4. Set two 3" scraps of the same molding in the corners on top of those spacers, sticking straight up, and staple them in place (A). Now set another two 3" scraps of molding against the 1×4, butted up against the spacers in the corner, and staple them in place (B) so that they hold the molding in the corners against the plywood sides.

Pull out the two spacers and flip the whole thing over. Set those same two ¾" molding spacers on the edge of the slightly sunken 1×4, butted up against the plywood. Set two more scraps of molding flush against them on the 1×4 and staple them in place (C). Pull out the two spacers. Now the jig will perfectly position the ¾" molding for the box lids when you slide them into that same spot for each new lid assembly.

To build the top piece of the jig, you will need one piece of 1×, two short strips of plywood (about 1" wide and just a few inches long), and one longer strip of plywood. It's important that the 1× is the same length and thickness as the 1×4s used for the boxes. The dimensions of the plywood strips aren't important.

Rip the 1× (D) down to about 1" wide to give better access for the stapler when assembling the lids. Staple the longer strip of plywood (E) to one of the 1" sides so that it overhangs one of the long edges. In use, this plywood strip sits against the end of the box lid and the end of the bottom jig. Staple the last two narrow strips of plywood (F) to the ends of the 1×. These will sit flat on the box lid, mimicking the bottoms of the sides of a box. When this jig sits on top of the end of a box lid, the ¼" molding piece can be properly positioned by sliding it between the two pieces of plywood (F) and up against the 1× of the jig (D), all while the jig is held flush with the end of the plywood lid.

To use the jigs, push two of the ¾" molding pieces used for the bottom cleats into the corners of the bottom jig, and run a thin bead of glue down each, being careful not to get any glue on the jig. Place a plywood lid piece on top of the bottom jig, lining it up flush with the corners and edges and being careful not to smear the glue too much. Place the top jig on top of the plywood lid piece, aligning the edges. Run a thin bead of glue on one of the ¼" molding pieces, and use the top jig to align that piece of molding on the plywood

lid piece. Staple each end of the molding. The staples will go through the plywood lid and into the ¾" molding pieces. Next, remove the top jig, flip the lid around in the bottom jig, replace the top jig, and glue and staple on the remaining ¼" molding piece.

If you've done this all properly, the lid should fit on a box with minimal effort, with the plywood flush against the top edge of the box. Then you'll be able to stack a second box on top of the first, and it will sit flush on the plywood lid, with little or no side-to-side movement possible between the boxes or the lid.

USING THE BOXES

The dimensions of these boxes take into consideration the bags I was putting into the boxes, the vehicle I was transporting the boxes with, and the ease with which I could carry a full box. You could easily change any of the dimensions to fit your own needs. If you plan to make boxes for a farmers' market display, you might want a much shallower design with only one 1×6 on the ends and a different handle. Instead of the lids, you might have locating dowels on the top corners, with matching holes in the bottom corners to allow the boxes to stack securely.

I wanted my boxes to be as lightweight as possible so that I wouldn't be transporting more weight than necessary. You could make this design even lighter by planing down the 1× pieces to ½-inch thickness. On the other end of the spectrum, if you're not so concerned with weight, or if you want significantly larger boxes, I recommend using ⅜-inch plywood for added stiffness. The boxes with lids, as built, weigh about 7 pounds.

Because all of the wood is left raw, the boxes are able to absorb a bit of water from the wet canvas bags. The lack of finish also means that

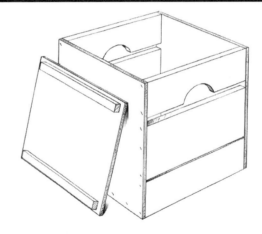

paint or stain won't wear off onto the bags themselves. I guessed that finishing the wood wouldn't result in the boxes lasting significantly longer, so by leaving the wood raw I was able to avoid extra economic and environmental costs. I did, however, print the Our Table Cooperative logo on each box using a custom silkscreen and a water-based ink fixer. It's easiest to print with a screen before the boxes are assembled. Another option is to get a custom branding iron, although that is more expensive than a silkscreen.

These boxes saw several years of heavy delivery use before the CSA adopted a different system. They were then retired to other, less wearing uses, such as displays in the farm store.

Tools Built for Women
A Fresh Look at Ergonomics

Ann Adams and Liz Brensinger started Green Heron Tools because as women farmers who also had master's degrees in nursing and public health, they knew what they were looking for in tools that fit women—and they weren't finding it in the marketplace.

With funding help from the USDA, they started working with scientists, engineers, and women farmers to identify exactly what their tools should look like and how they should be different. Ann and Liz looked at the way typical women work and found that because women have more of their strength in their lower bodies than in their upper bodies, they tend to favor working with their lower body muscles. Women also tend to have less muscle mass overall than men do.

The first Green Heron Tools, a shovel and spading fork, were for digging. To create a shovel that worked especially well for women, the science led them to incorporate a supportive tread on the blade to facilitate pushing with leg muscles and using body weight. They also modified the angles between the blade and the handle so that the blade more easily enters the ground at an angle, instead of straight down. This approach requires significantly less force in most conditions and mimics the way women in their studies actually worked.

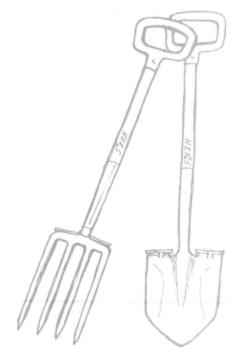

In addition, they designed a larger D-handle, making it easy to hold the handle with both hands, each one gripping at a different angle to maximize upper body muscle contribution when necessary. Ann and Liz also recognized that one size would not fit all, so they made multiple sizes to fit women of different heights.

These designs happen to work better for many men as well. A lot of this is common sense and is important to pay attention to with all tool design (science is largely the practice of paying very close attention to details and quantifying them). Some of the basic considerations for tool selection and design that Ann and Liz point out are:

- **Allow your joints to remain in a neutral (non-twisted) position.** One way you can do this is to set your work surfaces high enough so you don't have to bend over. I encourage you to take this into consideration when building many of the projects in this book. In addition, Ann and Liz point out how to keep your joints neutral while using a hoe. Pistol grips on hoe handles aren't common, but if the angle of the handle is tilted and intended to be used in a push/pull motion, which many are, having a pistol grip on the end of the handle keeps your wrist in a neutral position. Also, using an extra-long hoe handle allows a more upright stance, which keeps your back more neutral.

- **Appropriately size the griping surfaces—not too small, not too fat.** Consider this as you cut and round the handles of the CSA Boxes (page 150) and choose the pipe diameter and length for the crank handle on the Drip Winder (page 76).

- **Allow for adjustability if users of different sizes will be sharing the same tool.** The adjustable table height on the Rolling Packing Table (page 140) is an example of this. Ann and Liz also note that wheel hoes, like the one made by Valley Oak, have a quick-release handle adjustment that makes customizing the tool in the field quick and easy.

- **Keep tools light.** A heavier tool does not necessarily mean stronger or better. Making tools as light as is reasonable reduces the work needed to move and use them, which in turn reduces strain and potential injury. It also reduces the quantity of material needed, which can save money and resources.

CHAPTER 6
OFFICE TOOLS

TOOLS FOR PLANNING, RECORD KEEPING, AND ANALYZING your farm are just as important these days as any tool you use in the field or packing shed. The tools I talk about in this chapter push the edge of what most people might define as a farm tool, and the sections here are as much about creating good systems for using these tools as they are about building the tools themselves. The truth is that as a farm owner, I probably spend as much time using computer-based tools to plan, track, and manage the farm as I do using all of the other tools on the farm combined—unless you count my hands as a tool in the field. Good old pencil and paper are still favorites of mine, too, and they deserve a spot in any book on farm tools just as much as a sprinkler system or hand cart does.

SPREADSHEETS FOR SEASONAL PLANNING

Every year at some point you need to figure out what you're going to plant, when you're going to plant it, and where it's going to go on the farm. In addition, if you're working with farming partners, employees, or even volunteers, it's not enough to just figure out these details; you also need to relay the information to the people doing the actual work.

Some farmers may be good at doing this at the last minute, but most of us, especially when we're starting out, will benefit from making a detailed plan before the season even begins. This allows you to buy seeds and supplies before you actually need them, which results in two benefits: You get your seeds early (seed companies tend to sell out of popular varieties later in the season), and you can save money by knowing how much seed you'll need through the season instead of taking wild guesses. The more practice you have figuring out your planting details, the better you'll get at making related decisions.

My Process

FOR MORE THAN 20 YEARS NOW, I've used a basic three-step planning process that starts with my end goal and works backward to the present. This is basically the same method I use for designing tools: I start with the goals, work backward to plan a design, then figure out the materials and methods.

STEP 1: Create a harvest plan. Start by making a list of the crops you want to grow. For each of those crops, estimate when it will be first ready for harvest, how many weeks you plan to sell it, and how much of it you'll sell each of those weeks. With this information, and by knowing approximate prices for all of your crops, you can

easily project the gross sales for the year. You can then make a reasonable budget that keeps your expenses below the gross so that you'll have a little left over to pay yourself and reinvest in the farm. This might also be referred to as a sales plan or distribution plan.

STEP 2: Create a planting plan. Working backward from the sales plan, figure out how much of each crop you need to grow and when it needs to be planted to be ready to sell on your projected dates. For some crops, such as lettuce, this means planning many planting successions. For others, such as tomatoes, you'll probably just do one planting for the entire season.

	crop	Price	unit	22	29	6	13	20	27	3	10	17	24	31	7	Total Value
				June		July				Aug					Sept	
3	arugula	$9.00	lb	0.3	0.3											$ 5.40
4	basil	$10.00	lb		0.3				0.3			0.3				$ 9.00
5	beans	$5.25	lb			0.8	0.8	0.8	0.8							$ 16.80
6	beets	$2.60	lb	1.25				1.25						1.25		$ 9.75
7	cabbage	$3.00	ea	1												$ 3.00
8	carrots	$2.50	lb		1		1		1		1		1		1	$ 15.00
9	celery	$2.25	ea					0.5				0.5				$ 2.25
10	collards	$3.00	bu	1			1									$ 6.00
11	cucumbers	$2.35	lb		0.5	1	1	1	1	1	1	1	1	1	0.5	$ 23.50
12	eggplant	$2.50	lb					0.5		0.5		0.5		0.5		$ 5.00
13	favas	$3.00	lb		1.5	1.5										$ 9.00
14	fennel	$4.00	lb	1.5												$ 6.00
15	kale	$3.00	bu		1											$ 3.00
16	lettuce	$2.25	ea	1	1	1	1	1	1	1	1	1	1	1	1	$ 27.00
17	onion, breen	$2.00	bu	1												$ 2.00
18	onion, bulb	$2.00	lb			0.5		0.5			1		1		2	$ 10.00
19	potatoes	$2.00	lb			2	1.5			2				8		$ 27.00
20	sum. sq.	$2.35	lb			0.5	1	1	1	1	1	1	1	0.5	0.5	$ 19.98
21	sweet peppers	$4.00	lb							0.5	1	1	1	1	1	$ 18.00
22	swiss chard	$3.00	bu			1				1		1		1		$ 12.00
23	tomatoes	$2.75	lb					0.5	1	2	2	2	2.5	2.5	2	$ 39.88
24		share value		$ 22.20	$ 19.13	$ 22.48	$ 19.65	$ 19.15	$ 19.40	$ 20.70	$ 20.08	$ 23.70	$ 22.33	$ 40.15	$ 20.60	$ 269.55
25		# of items		7	7	8	7	9	7	7	8	8	7	9	7	

An example of a basic harvest plan spreadsheet for CSA shares

STEP 3: Create a planting map. Map out the planting plan so you know where on the farm everything will be planted. I have a relatively simple method that allows you to look at the map over time, so you can see where some of the beds on the farm can be planted more than once in a season.

Inevitably, you'll probably find that there isn't quite enough space for everything, or that it would be more convenient to plant some things in slightly larger quantities. The process is somewhat circular, with each step informing the others and requiring you to go back and make adjustments as you go.

Once you've completed all three steps and have a solid overall plan, you can pull out the information you need to make to-do lists for every aspect of the planting process. I make separate to-do lists for:

- Seed orders
- Field preparations
- Greenhouse seeding
- Direct seedings in the field
- Transplanting in the fields

My Tools

TO CREATE THE PLANS, I use a combination of records and data that are primarily on paper but increasingly are also digital. I create the plans themselves entirely in digital form using simple, customizable spreadsheets. The to-do lists, harvest plans, and maps all get printed on 8½ × 11-inch paper. These paper to-do lists, harvest plans, and maps also contain space for new

records, which become data that feed the plan in the following year or for midseason changes in the current year.

Over the years I've used different spreadsheet programs to make my plans. With a few minor exceptions they all work similarly, and many are commonly available (in some cases for free). Digital spreadsheets are incredibly powerful and flexible tools. Even with just a little understanding of some of their very basic features, you can save a lot of time in your planning process. For example, you can have the spreadsheet calculate quantities and dates for you or use the sorting features to rearrange your plan so that you can look at different aspects from different perspectives. In the next section I'll walk you through the features I rely on heavily and give a few tips that I think make planning with spreadsheets more approachable.

I currently use Microsoft's Excel for Mac, so my descriptions most closely match that program. In the past I've used OpenOffice, NeoOffice, Numbers, and Google Sheets, as well as a few others. Numbers is the only program I found to be significantly different in workflow and default features, but it can still perform all of the same functions I talk about here.

In the more than 20 years that I've been using spreadsheets for planning, there have been almost no new features introduced that I've found useful. I say this largely to point out that part of what makes spreadsheet programs useful is that they are relatively stable. If you put just a little bit of effort into learning some of their most basic features, they can be very powerful tools.

The Importance of Flexibility

I apply several engineering principles to my planning, including the concept of acceptable tolerances and variability. I always keep in mind that the plan is just my best guess, based on how things have worked out in the past for me or according to other trusted sources of information, such as cultural practice information in agriculture books and seed catalogs and tips from fellow farmers.

I know that weather and other factors vary from year to year, so it's impossible to predict exactly how much a planting will yield, exactly how long it will take to mature, or exactly when

we'll have time to do any of the specific tasks necessary to get from seed to harvest. Allow for wiggle room in all of your plans to accommodate the expected variability.

Functionally, there are two main ways I recommend building this wiggle room into your plans. First, use a 52-week calendar, with the Monday of each week signifying the entire week, and ignore all of the other days. Only specify the week in which you expect anything to happen and consider any day within that week "on time." In some cases you'll be off by a week, or even two, and with experience you'll get to know which crops you have more flexibility with and which need to stick more tightly to the schedule. You may even have a crop—such as baby salad mix or micro greens—that needs a slightly more rigid schedule, but those are exceptions. From a planning perspective you should have about a week of flexibility with the planting and harvesting schedule for almost all crops.

Second, recognize that crop yields will vary each year. For succession plantings, the yield will even vary from planting to planting within the same year. As an example, we can seed carrots the same way a few weeks apart—or even a year apart on the same day of the year—and we'll often get vastly different yields. The longer you farm, the more consistent you'll get with everything, and every year you'll also gain a better understanding of which crops are most predictable and which have more widely varying results.

Within the planning process I use my standard yield numbers (a likely average) to estimate how much space I need to plant for each crop to get the harvest I think I can sell. The reality is that calculated space estimate is almost never an even number of beds. Adjust the amount to make an even number of beds or a convenient fraction of a bed. For example, if you calculate that you'll need 68 feet of beans but your standard bed length is 75 feet, you'll probably just plant 75 feet of beans. Or maybe you really don't want to take care of, or pick, extra beans. If you have another crop that's compatible with beans and needs only 25 feet, you might decide to plant only 50 feet of beans, hoping for a good yield but knowing it's a bit of a gamble.

These kinds of judgment calls are required for every planting you'll plan—and they have to be made whether you plan in advance or make a decision at the last minute. Planning for them in advance is good practice for making the decision, no matter when you actually end up deciding. If you really need to, you can always change the plan at the last minute, which is a good thing to keep in mind if you're worried about being bound by your plans.

Creating the Harvest Plan

IF YOU'VE READ my previous book, *Compact Farms*, you've seen a simplified version of my harvest plan spreadsheet. As I alluded to earlier, this might be described more accurately as a sales plan or a distribution plan for the following reason: Across the top row of the spreadsheet, I list the distribution weeks, using one column per week. Down the first column I list all of the crops I plan on harvesting. Within this grid, for any particular week, I can put a number in the row for a certain crop; this number represents the units of that crop I plan to distribute that week.

The reason I call it the harvest plan, though, is because it tells me how much I ultimately will need to harvest in order to be able to distribute. In the spreadsheet I list the distribution weeks instead of the actual harvest weeks because it's easier for me to start by estimating how much I think I can sell in a week. If I were making the plan for my own garden, I would think about this sheet in terms of how much produce I wanted to eat each week.

In either case, it's important to note that I'm planning when I will distribute (by selling or by eating) the produce, not necessarily exactly when I'll harvest the produce. For example, storage onions or potatoes are harvested when they're ready, but I continue to distribute them over many weeks after they are harvested. So in the plan I list out the quantity per week of distribution, and then I simply add up those quantities to get the total harvest I need. The sheet could be as simple as this and still be useful, but adding just a few more layers makes it much, much more powerful.

> **TIP** FOLLOW ALONG
>
> If you're not very familiar with spreadsheets, the next couple of pages will probably make more sense if you follow the text while actually typing out what I'm describing in a spreadsheet program.
>
> Also, the internet is your friend! Spreadsheets are incredibly common, so it's really easy to look up answers to questions about how to use specific features. I do it all the time! If you're not familiar with any of the terms I use here, search for them online and you'll almost certainly find a more detailed explanation. For lack of space and time I'm leaving more a trail of breadcrumbs than detailed step-by-step directions.

Format Cells

It's worth paying some attention to formatting the cells of the spreadsheet to make them visually more informative. For example, bold the font in header rows to make them stand out, and add a line to the bottom of the header row to separate it from the data below. You can also freeze the header rows and columns so that they are always visible, even while you scroll to other parts of the sheet. This will make it much easier to read large spreadsheets.

Adjust the width of your columns so that they are just wide enough to show the data you need to see. For the harvest plan sheet make the first column (column A) the full name of each crop. Double clicking on the right border of the column header will automatically adjust the column width to fit the widest entry in that column, or you can click and drag the right border to customize the width. If you select multiple columns at once and drag only one of the column edges, the program will make all of the selected columns the same width.

Formulas for Dates

You can use a simple formula to fill in all of the Monday dates for the year without having to enter more than one actual Monday date. In the example harvest plan sheet below, make the first date entry in cell B2. Enter the Monday of the first week you plan on distributing—for example, 5/20/2030. (It's important to understand that every cell in your spreadsheet has an address. Cell B2 is the cell that is second to the right and the second down.)

In the next cell to the right, C2, type this simple formula: **=B2+7**. Once you hit return, cell C2 will display 5/27/2030 because it assumes that you wanted to add 7 days to the date in B2.

To check whether there is a formula in a cell, look in the formula bar. If there is a formula, you'll see its output—a date, in the previous example—displayed on the grid of the spreadsheet.

Copying Formulas

Copy cell C2, highlight as many cells as you like to the right, and paste the formula into all of these cells. For example, try highlighting cells D2 through Z2 and pasting cell C2 into all 23 cells. Each cell will show the next successive Monday, so at this point you have columns for 25 Mondays in a row.

If you look at the formula in each one of these cells, you'll notice that every one of them is different. This is because the "B2" in the first formula (=B2+7) is a *relative cell address*. When you put "B2" in a formula that is in cell C2, the program knows that B2 is one cell to the left and it understands that it is supposed to reference that cell. When you copy cell C2 to D2, the spreadsheet translates it to "=C2+7." This makes it easy to write a formula that can be copied and pasted throughout the spreadsheet—but you do need to be careful about how you move formulas. More on that later.

Using a formula to calculate dates is an easy way to set up a spreadsheet to display all the Mondays in a given time period. You can easily update the spreadsheet each year by simply changing the Monday in the first cell.

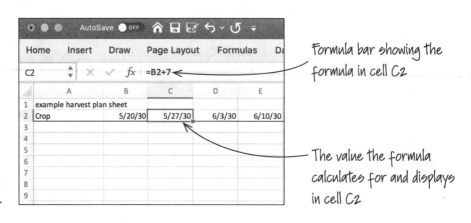

Formula bar showing the formula in cell C2

The value the formula calculates for and displays in cell C2

Tricks for Formatting Monday Dates

You can format the date cells so that they show only the day—and not the month and year—allowing you to make your columns quite narrow. To keep track of the months, manually type in the month names in row A above the first Monday in each month (displayed in row B). I find this easier to look at than a string of cells with longer date formats, which require wider columns. To format the dates this way, under "Format Cells," click "custom" and type *dd*. Alternatively, you can format the dates in any way you find easy to read.

Some people like to add color fill behind certain types of cells. I'm not much for colorful spreadsheets, but I do use a light gray fill to signify that a cell has a formula in it and shouldn't be altered.

Additional Data Fields

I recommend adding a few more columns of data to your sheet that should prove helpful in your planning. After the crop column, insert two columns before your week columns begin. In column B add a column header for average price per unit, then format that column as currency.

In the second new column (column C), specify the unit for each crop—for example, *lbs* (pounds), *ea* (each), *bu* (bunch), or whatever makes sense. Having a column for the unit designation means you don't have to include a unit in the price column or in the week columns, where you'll put the quantity for each week. With only numbers (no text) in those columns, you can use those values in formulas. If you include the unit in the same cell as the number, your spreadsheet won't recognize the number as a number and therefore won't be able to make calculations with it.

On the far right side of the sheet, just after your final harvest week, add a column for the total units for each crop you plan to sell that year. To add together the units from each week, use the "sum" feature. For example, let's say your first crop is arugula and it's in row 3. If you've set up your spreadsheet how I've described,

the weeks should start in column D and go through column AB (25 weeks of harvest). In the total units column (AC), insert this formula: **=sum(D3:AB3)**. The spreadsheet will add up all of the numbers in the cells between D3 and AB3.

In the next column (AD), you may find it helpful to remind yourself of the units used for each crop. Instead of copying and pasting each cell from column C, you can write a very simple formula, **=C3**, to automatically copy the unit from cell C3 into cell AD3. Then copy cell AD3 and paste it into the cells below in column AD. If you change the units of a crop at a later point, remember to do it in column C, not in column AD. The new unit will automatically update in AD. (This is why I add a light gray fill in formula cells: to remind myself not to make updates to numbers or text directly in those cells.)

If you want to figure out how many dollars' worth of produce that represents, use this formula in the next column (AE): **=B3*AC3**, assuming column B is where you have listed the price for each of your crops and column AC is where you have totaled the quantity you plan to harvest. Again, once you've written this formula you can simply copy cell AE3 and paste it down the column for all the crops you have listed, and you'll get the totals for each crop.

You can figure out your gross sales projections with a simple formula, too. If you've set up your spreadsheet as I describe, insert this formula in cell AE1: **=sum(AE3:AE1000)**. Assuming you've listed fewer than 1,000 crops, you will now have an estimate of how much your farm will gross for the year. All of these formulas will update automatically as you make changes to your numbers in the plan. Very exciting stuff!

Of course, you can always include other data fields, and there are many variations on the setup I describe here, but this should get you started. If you never add any other layers to your spreadsheet, you'll still have a ton of good information and you'll be more than ready for the next step.

CUSTOMIZE YOUR PLANS TO FIT YOUR NEEDS

If you're using the harvest plan sheet to plan out CSA shares, make a plan for a single share. Then you can multiply by the total number of shares you plan to sell to get your total harvest quantity for a given week. (I usually do this multiplication in the planting plan itself.) This allows you to see easily what a single share consists of. You can then easily adjust the total amount needed of each item if you decide to change the number of CSA shares you are offering.

If you plan on selling both CSA shares and to restaurant accounts or at a farmers' market, make two separate harvest plans on separate tabs of your spreadsheet workbook. If you use the same list of crops and the same list of dates on both sheets, you can use a third tab to easily combine the two, writing a formula that will add the quantities from each of the two other sheets. To add together the values from the two other tabs, simply write a formula that references the cells in the other sheets. (You can search online for "combine data from multiple worksheets" to figure out how to write your formula.)

The easiest way to insert a cell address into most formulas is to start typing the formula with an = sign. When you need to enter a cell address into the formula, click on the cell you want (even if it's in another tab) or use the arrow to move to the cell you want to select. The cell address will automatically appear in the formula, and you can continue typing the next part of your formula.

NOTE: Cell addresses from other tabs begin with the name of the tab, followed by an exclamation mark and the cell address within the tab. For example, cell "csashare!G25" is found on the tab named "csashare," and the selected cell in that tab is G25.

Instead of selecting a single cell for a formula, you can also select a range of cells, which will show the addresses for the top left cell and the bottom right cell separated by a colon. This is useful for functions such as "sum," which adds up all of the values in a range of cells.

When you are adding the totals from the restaurant and CSA share tabs together, you will need to multiply the individual share amount by the total number of shares offered. You can designate the total number of shares in a single cell at the top of the sheet. When writing the formula, you can use an "absolute" cell address (as opposed to the default "relative" cell address) when referencing the number of shares. This allows you to copy the formula across rows and columns without the cell address for the number of shares changing.

Unlike a relative cell address, which will change as you copy your formula into new cells, an absolute cell address stays the same no matter where you move the formula. To make the column absolute, add a $ sign before the column name ($A1). You can also make a row absolute (A$1), or you can make both a row and a column absolute (A1).

Creating the Planting Plan

BEFORE YOU DIVE INTO the next steps, I have two tips for you. First, think broadly when you're creating the harvest plan and don't get overly specific. For example, I recommend simply listing "head lettuce" rather than including every kind of head lettuce you might grow and sell for more or less the same price. You can separate out red leaf, green leaf, butter lettuce, and so on in the next step of the process.

Second, remember that these are very rough estimates. Even if you manage to grow everything you include in your harvest plan, you also have to sell all of it in order to make the big gross number in cell AE1. Spreadsheets give you nice, precise numbers (with as many decimal places as you tell them to), but that doesn't mean you'll get all of that precision out of the guesses you input. The calculations the spreadsheet makes are only as good as the numbers you feed it.

Start an entirely new spreadsheet to create your planting plan, which will be based entirely on your harvest plan. This is where you will detail every planting you'll need in order to have enough produce on the right weeks for your harvest plan to work. You can separate crops by varieties, and if you want to experiment with different plant spacings or different planting dates, include those here, too. Think about every question someone working on your farm might ask when heading out to the field to do a planting, then answer it in this spreadsheet as concisely as possible.

I think of my planting plan spreadsheet as the master plan. Within it, I include a few formulas to help me make decisions as I fill in the details for these plantings. I explain some of these formulas below.

Organize the Plan into Sections

To help you keep track of all the information on the sheet, roughly group the columns into sections. Start with the two simplest pieces of information:

a column for the crop type and a column for the specific crop variety. For example, "kale" is a general crop type, and "Lacinato Rainbow" is a specific variety of kale. If crop sub-types are needed, I put those with the general crop type, using a comma. For example, "Lettuce, romaine"; "Lettuce, butter"; and "Lettuce, baby leaf mix." This helps with sorting by crop and variety later.

Harvest plan and estimates. After the crop and variety columns, carry over information from your marketing plan to remind yourself when and how much you hope to harvest from a specific planting. Create a corresponding section titled "Estimates" with columns listing approximately how many weeks each crop variety will take to mature from planting to harvest and how much each variety typically yields for a given space. With those numbers, you can write formulas that will give you estimates for planting dates and the amount of space to plant.

For example, one formula I use takes my expected first harvest date in column C and subtracts the expected number of weeks to maturity for the crop in column K (multiplied by 7 to get the number of days) to give me a suggested planting date. For a planting listed on row 15 of my spreadsheet, the formula would look like this: **=C15−(K15*7)**.

Note that I use parentheses here to make it explicit that the K15 should be multiplied by 7 before being subtracted from C15. Using parentheses isn't actually necessary in this particular situation, but it does make it easier to understand if you need to edit the formula later. (Search online for "order of operations in spreadsheet formulas" if you're curious to learn more.)

Planting plan. In this section of columns, use the information from the Estimates section to guide your decisions and enter the week you *actually* want to plant each crop, as well as the amount of space you *actually* plan to plant for each crop. Include columns for not only the number of beds (or fractions of a bed) but also the number of rows you plan to plant on the beds, your desired in-row spacing between plants, and whether you plan to direct seed (ds) or transplant (tp) the crop.

All of my fields are labeled with a letter, and beds are labeled with a number, so in this section I include the location where I expect to plant the crop (the exact locations will be determined in the planting map; see page 172). I also add any special notes, such as "plant in trenches," to remind me and/or my crew how to plant that crop if it includes any practice that's a little out of the ordinary.

Greenhouse. In this section, collect information and make calculations for any crop you plan to grow your own seedlings for and transplant. On our farm, because we grow all of our own seedlings in a greenhouse, we include in this section the number of weeks we think the seedlings will need in the greenhouse, what size of plug tray we should seed them in, and how many seeds go in the tray. I use these details in combination with the information from the planting plan to write formulas to determine what week to start the seedlings and to decide how many trays to seed.

For example, to calculate the minimum number of plants I need to transplant into the field, I write a formula that references the number of bed feet I plan to plant (column Q), the number of rows I plan to plant in the bed (column S), and the in-line spacing between the plants, in inches (column T). The formula for the number of plants I'll need of the crop listed in row 6 is: **=Q6*S6*12/T6**.

Direct seeding. Add a few columns for crops that you seed directly into the field instead of transplanting. You can include seeder settings for each crop in these columns so that you'll have them when you head out to plant.

Seed order. This final section starts with some basic data: average number of seeds per ounce for the crop type. Some seeds are sold by the count, some by pounds, some by ounces, and some by grams, so include a separate column for each of these calculations. For example, there are typically 19,800 carrot seeds in an ounce, but I sometimes buy carrot seeds that are sold by the count instead of by weight. Having both numbers handy makes it easier for me to decide how much seed to order.

In addition, you might find it helpful to have columns where you can remind yourself if you have any seed that is left over from the previous

TIP GERMINATION RATES

One factor I consider when seeding in the greenhouse is legal minimum germination. Seed companies are required to conduct germination tests on their seed lots, and in order to sell the seeds, those germination tests must yield above a minimum number. Each seed type has a different minimum, generally ranging from 55 percent to 80 percent, which you can easily look up. Most seed companies that work with commercial growers sell seeds well above the legal minimum germinations.

On very rare occasions I have gotten a nearly 99-percent germination rate, but usually it's significantly lower, so I over-seed in order to have enough good plants. Generally I expect to get at least the legal minimum germination, so I use that number to guide my decision on how many plugs to seed.

year and still usable, what seed company you need to order from, the quantity you're ordering, and the cost for the seed.

Sorting

Both the marketing plan and planting plan include a lot of information. I commonly have more than 150 distinct plantings in a year, and when I have worked on especially diverse farms, my planting plan has included more than 400 plantings! With such large amounts of information, sorting is a very powerful tool.

One place to use the sorting function is when your planting plan is complete and you want to make your seed order. Sort all of the rows by the column containing the seed company name. You can also have your spreadsheet run multiple levels of sorting at the same time. For example, you can sort by crop name, crop variety, and quantity to order. This lets you organize how much of each type of seed you need to order from each company.

In Excel, to add multiple levels of sorting, highlight the rows you want to sort (I find it easiest to select all of the columns in the rows by clicking on the row header), then select Data, then Sort. In the window that pops up, you can specify the columns to sort by (in ascending or descending order), and add sorting levels by clicking on the plus symbol. (I don't include any header rows and don't use the "my list has headers" option, which the program sometimes defaults to.) Other programs have similar multilevel sorting functions; search for tutorials online.

You can also sort the sheet by crop, by planting date, or by harvest date. Being able to sort quickly in multiple ways will save you a ton of time in your planning process before you even get to the seed order. (See Using the Plans on page 174 for more examples of how to use the sort function.)

WARNING: When you sort a spreadsheet be sure to select *all* of the columns in the rows you are sorting that contain information. If you don't do this, you can easily disassociate information within a row in a completely irreversible way. To ensure that you sort all of the information in the columns, use the row header to select the full row instead of selecting only a few columns in the row (unless you intentionally want to separate some of the columns from the others). Save your work frequently, and always before you do a big sort. That way, if something gets screwed up during the sort, you can go back to the last saved version.

Be aware that sorting can also change the way some formulas reference other cells. This usually works in your favor, but not always, so double-check your formulas after sorting. (See Using the Plans on page 174.)

Creating a Planting Map

IF YOU LISTED in your planting plan the field and bed location where you want to plant each crop, you can make sure it all fits together by creating a map using a spreadsheet grid. If you skipped that step, now is the time to decide where you will plant everything. Start an entirely new spreadsheet for this map.

On the farm I currently manage, Cully Neighborhood Farm, there are seven fields, labeled A through G, and each field has 10 beds. For each field, I make a separate map that fits on a single piece of paper when it's printed out. These maps are all in the same spreadsheet "workbook," and I use a separate tab (or sheet) within the workbook for each field, plus one blank map as a template.

In the first column I designate rows 3–12 each as a separate bed. Across the top (starting in cell B2), I use the same formula that I used in the harvest plan sheet to mark out the Monday dates. I include each Monday of the year, for a total of 52 columns of weeks. In this spreadsheet, the rows designate space and the columns designate time, so I often refer to this as a map in space and time.

Once you have the basic template you can then "plant out" your entire farm on paper by transferring every planting line from the planting plan to the map. This is a very condensed map, so I recommend using a lot of shorthand. For each planting, type the variety name into the cell that is in the correct row and under the correct week. Add an *H* to the cell(s) on that same row that correspond to the week(s) that you plan to harvest from that planting. I fill in each cell from the planting week to the last harvest week with a light gray so I can easily see when beds will be occupied and when they will be free. The cells that are white on the spreadsheet represent weeks when beds are available either for other cash crops or for cover crops that will help feed the soil.

Inevitably, as you transfer over data from the planting plan to the maps, you'll realize that either you don't have enough space in some areas or you have available space in some areas. At that point you'll need to decide what changes you want to make to the harvest plan and/or planting plan to adjust to the reality of how much space is available at any particular time. This becomes a bit of a circular process, but with a little back and forth and careful consideration, you'll arrive at a good plan.

TIP GET THE MOST FROM YOUR SPREADSHEETS

Spreadsheets are among my favorite, and certainly most-used, tools on the farm. It takes a little bit of practice and knowledge to use the spreadsheet software, but once you're comfortable with it, you can easily customize your spreadsheet, pull out data, and write new formulas to perform automatic calculations. Because there is so much data in these sheets, and so much of the work involved with getting information into and out of the sheets is somewhat repetitive, it's important to understand how the sheets work and to establish a good workflow to streamline how you use them.

Over the years I've seen an increasing number of computer applications that attempt to help with crop planning and record keeping. Initially they all oversimplified, or only did small parts of what my sheets would do. These apps continue to improve, though, and I can see a day where they may have enough advantages to compel me to switch over.

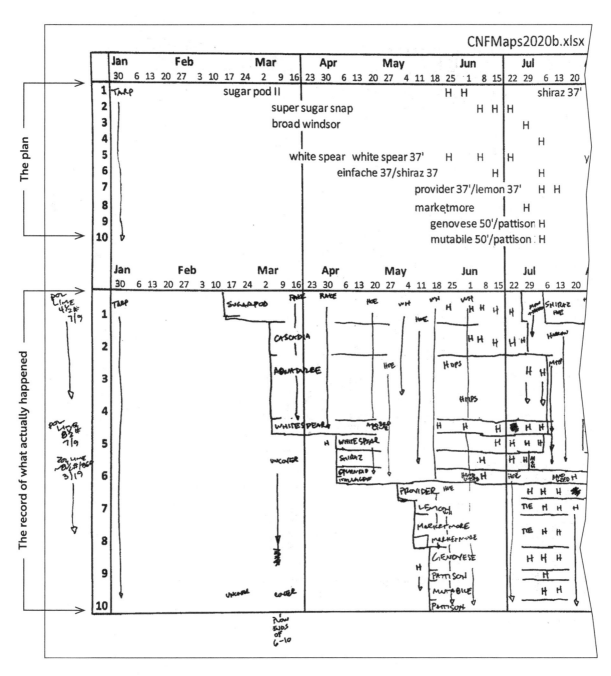

You'll notice that I stack two versions of the map on one sheet. The top one is the plan. The bottom one is a blank version that I use to handwrite notes of what actually happens during the season. This is the single most useful record sheet I have. It helps me understand how many weeks a crop takes to reach maturity for a particular planting date as well as how many weeks we typically harvest from a single planting of any given crop, among other things.

USING THE PLANS: CREATE TO-DO LISTS THAT ARE ALSO RECORD SHEETS

When you feel like you have a good plan, there's enough space to plant everything you want to, and the income projections are good enough to cover your expenses, it's time to make your seed orders and print out your plan on paper.

Both the harvest plan and the planting plan are way too big to print out on a single sheet of paper in a form that is easy to look at and use. Instead, break up those plans into usable sections that will fit on single sheets of 8½ × 11-inch paper. By fitting each entry on a single line, these printouts can become to-do lists.

From the planting plan, I print out separate sheets titled Greenhouse Seeding, Bed Preparation, Planting, and Direct Seeding. Each sheet includes a full list of what needs to be done, all of the relevant information for each specific task from the master plan, and blank columns to be filled in when tasks are completed. These lists are sorted chronologically to provide week-by-week instructions. They are also sorted by other important groupings so that all of the plantings in one field on a particular week are together, for example, or all of the seeds going into one tray size in the greenhouse are listed together.

My Greenhouse Seeding to-do list starts with the date column, followed by a blank column for the actual date when we do the seeding. When we write the date in that column, we're essentially checking off that the task is done—and keeping a record of when it was done.

The sheet also tells us what crop and variety to seed; indicates the number of plants we ultimately want; suggests how many to seed; and specifies the tray size, the number of trays to seed, and the expected field planting date, along with any additional notes. If we do anything differently from what was planned, we make a pencil line through the printed plan—leaving the original text visible, but indicating that something changed—and then write in what actually happened.

The primary copy of the Greenhouse Seeding sheet lives in the greenhouse so that it's easy to reference and fill in. This approach creates a complete record that helps answer questions that may come up when checking germination the week after seeding, when pulling trays of mature seedlings to plant out in the field, or when (at the end of the year) I want to update the plan and make improvements. At the end of the year, I create a file for that year's records, which all fit in a compact space and are in a form that's easy to reference any time in the future.

USEFUL AND USABLE DATA COLLECTION SHEETS

I TALK TO A LOT OF NEW FARMERS who want to know what tools they can buy to make their farm work, or to work better. The reality is that before you can know whether a tool is going to be useful for your farm, you need a little data. Farming is a business and, for many, also a vocation. The business owner in me says you need the data to run the numbers: Will a tool pay for itself? For the agriculturalist in me, the finances don't matter as much as figuring out what tool will help me produce the best product and be pleasant to use.

These two perspectives aren't mutually exclusive, and they both benefit from a bit of data collection and analysis. If you're someone who prefers "gut" or "intuitive" decision-making, think of collecting the data and doing a little analysis as practice that helps you improve your gut reactions and your intuition.

My Process

Before I start collecting data, I identify the question that I want to answer, and then I create a list of all the things I need to know—the data—to answer that question. Next, I consider what task I'll be doing when I record the data, and whether I can record it with minimal effort. It's also helpful if I can give myself physical cues to remind me to record the data. Sometimes I can automate the data collection, as I explain in Hacking Photos and Video for Data on page 178.

In addition, I want it to be easy to find and retrieve the data later and to make it simple to interpret. Usually, this means creating a recording sheet that will always be with me when and where I need to collect the data. The sheet should include

explicit cues for me (or whoever is filling it out) detailing everything that needs to be recorded, along with a visual example of how to record it properly. The greenhouse seeding sheets that live in our greenhouse (see the facing page) are a great example of this type of record sheet. The harvest record sheets that live in our packing area are another record sheet we use all the time.

My Tools

I'm starting to use digital methods to record data in some specific circumstances, but for almost everything I still think paper and pencil (usually used with a clipboard) are hard to beat. Pencils are inexpensive, they don't fade with time, and

Date: 6/24		CSA Count: 38 + 1	PLAN			ACTUAL			
start	finish	crop, variety	count (units)	weight (units)	field/ bed	count (units)	weight (units)	field/ bed	notes
JV 9:05	JV 9:25	BROWN GOLDRING	39@			39@ 4@ 4@	Brown G Nevada San	G5	BG Elongating
CCTM 8:35	9:39	KALE	39 BU			39@	17.9#		
JV 9:29	JV 9:33	BROCCOLI	ALL			18@	2#	D8-9	
CCTM 9:41	10:05	BEETS - SHIRAZ	2/3 BED		BB	121@	33.1#		
10:09	11:14	TURNIPS	2/3 BED		G6	200@	41.55#		4.9 17.9 18.75
TM 11:00	TM 12:20	SNAP PEAS	ALL		B10		17.3#		
CC 12:00 TM 1:15	CC 12:45 TM 2:00	FAVA	1 ROW+		D11		48.3#	D11 S	19.3# (½) 29.0#
JV 9:36	JV 9:50	SQUASH	ALL		B2-3	35@ 14@	8.2# 2.6#	B3 B2	Genovese (light) mutabile (dark)
TM 12:20	TM 12:38	GREEN ONIONS	39 BU			78@	8.4#		
JV 10:15	X	BASIL	1 LINE			88@	3⁹/4#	HH	
TM 10:05	TM 10:27	YELLOW BEETS				80@	28.1#	B6	

An example of a harvest record sheet. Not every record is perfect, but there's a ton of useful information that gets used on the day of the harvest and later in the year when I make crop plans and analyze production costs. Cross-referencing this sheet with my maps gives me even more information.

they even work when damp. We have a pencil sharpener, but everyone on my crew also has a knife and can sharpen a pencil with it. Paper is great because it is relatively inexpensive and surprisingly durable. We don't always keep the paper completely free of dirt and water, but with a little care we have no problem keeping it clean enough to survive a day of harvest in good enough shape that it can be filed and referenced later.

To create my record sheets, I make a simple grid using a spreadsheet program such as Excel. I include a separate column for every piece of information I want to record and combine to-do sheets with record sheets. For example, at the top of my harvest sheets I have a specific location to record the date and the number of CSA shares we're harvesting that day. I include separate columns for the crop/variety; the amount we plan to harvest by count; the number of pounds we plan to harvest (if applicable); and the specific location in the field where we expect to harvest that crop.

When someone on the harvest crew heads out to the field, they put their initials and the time in the start column. When they come back from harvesting that crop, they write their initials and record the time they got back. There are also columns for recording how much they actually harvested (in unit count or by weight) and where the harvest came from.

We print out these sheets using an inkjet printer, so the print does bleed a bit if it gets wet. I prefer laser printers or photocopies since they hold up better in damp conditions, but on our tiny farm we print so few sheets each season that I haven't bothered to replace my old inkjet.

It's possible that overall we spend too much time recording what we do on the farm, but I often find the information useful in planning and evaluating changes. I also think that paying close enough attention to recording what we do accelerates my learning and the learning of my crew members.

Building Record Sheets

Build your record sheets so that everything fits on a standard 8½ × 11-inch sheet of paper. Sometimes this means you may have to use the portrait orientation instead of landscape, but you shouldn't need to go any wider. Try to format the sheets so that they're pleasant to look at and have spaces that are large enough for someone to reasonably write the required record but small enough that you can fit a good amount of data on the sheet before needing another piece of paper.

When I fill out the record sheets, I often use some shorthand, but I try to make the meaning obvious enough that everyone on the crew can understand. Sometimes I also print reminders of standard shorthand in the header or footer, which helps if I have to come back to a sheet several weeks, months, or years later.

Even if I'm the one who originally filled in a sheet, it's much easier to analyze the data if everything is obvious and I don't need to spend time interpreting something vague. This is doubly true if someone on my crew has to refer back to a sheet; I don't want them wondering what I'm asking for or what information they're supposed to enter.

TIP HANDY DIGITAL HELPERS

In some sense, my reliance on paper records has a few downsides. Perhaps most noticeably, record keeping takes time. It also takes time to transfer the handwritten records into a spreadsheet. It would be nice if we could add data directly into spreadsheets, but in practice I've found it's faster to record the information on paper first and then transfer those notes into an electronic form later. Doing it this way lets me refer back to the paper frequently during the day, which I find faster than taking out my phone, opening the right app, and scrolling to the record I need.

That said, I have started to use some simple programs on my smartphone to augment my paper record keeping. In the Notes function on my phone, I have a specific folder and format that I use for weekly to-do lists. I create each list on my weekly field walk, and I update it throughout the week. Having the list on my phone allows me to share these notes with other members of the crew. I also use my phone to set alarms to remind me to do critical tasks, such as flame a bed of carrots or change the harvest day for a CSA member on a particular week.

Another digital tool I've started using is Google Forms to enter time studies of common tasks. That feeds data directly into a spreadsheet where I can access the information later to analyze the labor costs for those tasks and future labor needs.

HACKING PHOTOS AND VIDEO FOR DATA

I OFTEN WANT TO KNOW how long it takes to do something. For example, we pay for metered water and want to use it efficiently. Part of using it efficiently is knowing how long it takes to irrigate and how much water is getting to the fields in that time. We use the information from previous irrigation runs to inform the next irrigations, including when and for how long they need to run based on weather, plant responses, and soil core sample.

Another example is knowing how long it takes to do a certain task, such as planting a bed of lettuce. We use that information to help schedule future plantings of lettuce or similar crops. In addition, that data helps us calculate how much the lettuce costs to produce.

In the past, recording time meant I needed to start a stopwatch or note the times when I started and finished a task and then store that information in a place where I would remember to look when I needed it. Then, a few years ago, I realized that in my smartphone I carry a camera with me everywhere I go, and that it's much easier to take a photo that captures the important information than it is to jot down individual notes.

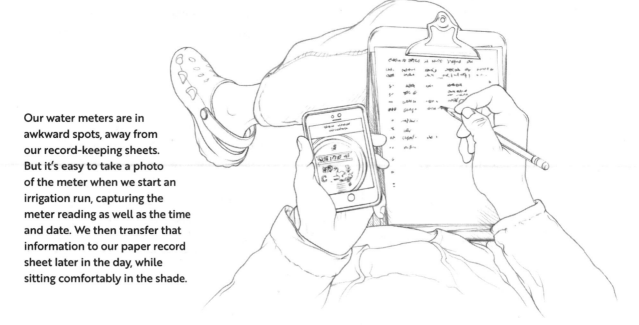

Our water meters are in awkward spots, away from our record-keeping sheets. But it's easy to take a photo of the meter when we start an irrigation run, capturing the meter reading as well as the time and date. We then transfer that information to our paper record sheet later in the day, while sitting comfortably in the shade.

USING PHOTOS FOR DATA

Before I used a camera for record keeping, I mostly wrote things down. In the irrigation example, this meant I had to note the meter reading, date, and time at the start of the irrigation and again at the end. For the planting example, I needed to write the date and time as well as the amount of space and number of people involved in the planting of a particular crop.

For both of these examples, it's easiest if I can quickly take the record in the moment and then, once a day (or once a week), move all of my records into a common file where I can easily find them later. Each snapshot with my phone makes an excellent initial record.

With the irrigation example, I take a photo of the meter when I start every irrigation run. This not only captures the meter reading but also the date and time of each photo. For a complete record I can take a second photo immediately after irrigating. We typically run about eight different irrigation sets per day. At the end of the day I transfer all of the records to an ongoing paper log.

To record something like a planting, I take a photo at the beginning of the planting that shows the bed we're planting into and the workers doing the planting. I take another photo at the end of the planting. We always keep the record of the quantity planted in a paper record sheet, but we can use the time stamps on the start and end photos to know how much time it took to plant, as well as the exact date and time something was planted. I find this slightly faster than accessing my phone's stopwatch, which ultimately only gives me the elapsed time.

USING VIDEO FOR DATA

A time-lapse camera is handy for doing time studies like the one I mention above for planting, or really for any task. The video created by the camera is easy to review quickly and to pull times from, assuming you know the ratio between the video speed and real time. For example, a $2\times$ rate would let you review 30 minutes of work in 15 minutes. A $60\times$ rate might be the most useful, letting you review 30 minutes of work in 30 seconds.

This could be especially helpful with tasks like planting where more than one person is involved, but not everyone works from the start to the end. With a time-lapse video, or a camera that takes one shot every minute, you could go back and review the task and just multiply the time by 60 (the camera speed factor) to find the actual time.

One way that I have used video is to do a weekly video "tour" of the farm. I slowly walk by each field while panning with the camera and narrating key points. I do this mostly for our farm's CSA members so they can see what's happening on the farm, but it also doubles as a future record of what fields looked like on a particular week, what stage plants were at, where we had weeds, and where fields were clean.

BUILDING RECORDS FROM THE DATA

Building a record is a matter of creating the record-keeping habit as part of every task that benefits from having a record. Start by establishing a system that will work consistently and quickly. The second, equally important, part is remembering to record the information. Finally, check that the record is complete, and file it in the proper place so that you can find it days, weeks, months, or years later.

I find myself using many of these records within a week or two of making them to help me answer questions as I evaluate the progress of our crops. Our irrigation schedule is very tight, and knowing when the last time a crop was watered helps me predict when it will need water again or understand problems that might appear in the crop. Having the irrigation records also helps remind me every spring when

irrigation season starts how long to run an irrigation set.

For me it's quicker to locate, reference, and read through paper records than most computer files. I keep all of my records for a year in a simple binder, and they usually don't stack up to more than about ¼ inch. I can pull out the binder for a particular year and quickly flip through the pages to find the original record.

I transfer parts of some records to spreadsheets, where I can analyze the data. For example, I have spreadsheets where I record the start and stop times (found in the photos) for time studies. This lets me see how the numbers change over time. We don't take these kinds of records every year for every crop, but we check back with them occasionally to see how we're doing.

DESIGN NOTES

One more way that I use photos as a record is particularly relevant to this book: as an aid to designing and repairing tools. I constantly check out farm tools on other people's farms and take pictures of details I want to remind myself of later. Similarly, if I need to disassemble something complex, I take a photo first so I will know how it goes back together.

When building a tool, I also take pictures at each step in the process if I think I might build it again someday. That photo record helps me remember the process so that I

can efficiently construct another version, even years later. In fact, many of the illustrations in this book are based on photos I took for myself over the years to record how I put something together the first time. Throughout writing this book I constantly looked at old photos, searching for the details of how I actually put together the Mini Barrel Washer, the Rolling Packing Table, and pretty much every other project in these pages.

NOW, MAKE THE DESIGNS YOURS

DESIGNING AND BUILDING TOOLS is something I do because I love experimenting and creating, and new ideas constantly come to me. When I say "new ideas," I don't mean that they are new in the world, but they are new to me, and I want to explore them in a hands-on way. Part of why I've written this book is to share with you ideas that I have gotten from other places, found interesting and useful, and think you might find interesting and useful, too.

The idea of protecting any of "my" ideas as intellectual property doesn't appeal to me. That's partly because I'm not sure how novel any of my ideas have ever been, but also because I just love sharing great ideas that I've seen out there—and the variations on them that I've created.

All of my ideas are clearly influenced by other people's designs. All of my designs are made from commonly available materials. I do put a lot of thought and work into improving my designs, but I'd rather put those thoughts and physical tools out into the world as part of a discussion about how tools are made and how we work than to try to hold on to any one of them as my own, controlling how and where it is made and used.

I hope you'll take these designs as inspiration for creating your own tools. And when you create your own tools, I hope you'll share your designs and tools with others—even me!—and in the process help us all better understand the myriad possibilities and variations that continue to move the world of farming and tool design forward.

APPENDIX

Designing Your Own Tools

I HOPE THAT YOU'LL BUILD THE TOOLS in the previous pages, at least the tools that seem like they'd work well for you, that you'll learn something from reading about the others, and that you'll adjust and modify the designs to make each tool work best for your particular situation. You probably have ideas for tools you'd like that are completely missing from this book. The upcoming sections titled Mechanical Principles in Plain English (page 191) and Materials Properties in Plain English (page 198) will give you good background information on those topics. There is also a basic process that I use for creating a design to make sure I'm taking all of the factors into consideration. I'll outline that basic process here.

Design Considerations

Considering the need. Early in my design process I think about what I need a tool to accomplish and how, specifically, I will use the tool. I envision myself doing the necessary task with the tool and try to imagine each step in the process. I think about what would annoy me or get in the way of completing the task efficiently, and then I try to incorporate solutions to those potential problems into my tool design.

For example, I know from experience that I prefer to not have seedling trays hanging off the edges of tables, where they might get bumped. So when I designed the Simple Seedling Bench (page 26), I made it wide enough to comfortably accommodate two rows of the 20-inch seedling trays we use in our greenhouse. I also measured my own hips and spaced the benches far enough apart in the greenhouse that I wouldn't bump into them as I walk down the aisles.

Similarly, when I began designing the Simple Spray Table (page 120), I realized that for years I had cleaned produce while standing in mud puddles trying to ignore how uncomfortable I was and how much it drained my energy. To prevent that same situation, I incorporated a piece of metal roofing into my table design to divert water and soil away from the user's legs and feet. This makes it much more comfortable to work at the spray table, especially in the colder months.

Deciding on dimensions. I offer dimensions for each project in this book, but before you build your own version, think about whether those dimensions suit you. Consider where you will use the tool. Do you need a smaller version of the tool so that it will fit in your space and give you enough room to move around comfortably? I designed the Mini Barrel Washer (page 128) so that it would fit inside a particularly small shed, but you may have room for a larger version—or need to modify my dimensions to make an even more compact version.

In addition to considering the physical space where I will use a tool, I also consider standard dimensions for lumber and other materials. For the Simple Seedling Bench, for example, I knew I would build the benchtops out of wood lath, which comes exactly 4 feet long. I saved time and materials in assembly by making the benches 8 feet long, or two lengths of lath; 8 feet long also happens to be a comfortable length for two people to carry when tables need to be moved or rearranged as they sometimes do.

Last but not least, when determining a tool's dimensions I consider the dimensions of the person (or people) who will most frequently use the tool. This is especially important when building tables, benches, or other equipment that you'll stand at for extended periods of time. Repeatedly bending and reaching can strain your body. If you have the chance to customize the height and/or width of a tool to accommodate your height and arm span, take advantage of the opportunity. I know that I can reach about 2 feet without having to strain at all, so I designed the Simple Spray Table to be 2 feet deep by 4 feet wide so that I don't have to move very much from side to side to reach the edges.

Planning for fatigue and failure. For each design, I also think about what forces the tool will see in both normal use and the reality of everyday life. A table will have things placed on it, so you might just design a table to accommodate vertical forces, but in reality a table experiences sideway forces, too, every time it gets leaned on and shoved from place to place. When planning my design, I think about not only the first time the table will get pushed sideways, but also what will happen the hundredth or thousandth time it gets leaned on.

Repeated stresses cause fatigue and gradually weaken parts and joints. Wear to surfaces reduces the thickness of parts, which in turn increases the stress on those parts. As I consider the most likely place for a design to break, I try to imagine the least critical place for the break to happen as well as the easiest place to repair a break. I try to design my tools so that if they do fail, the failure won't be dangerous or catastrophic. I also design them to be repairable with relative ease.

Design Constraints

It's quite possible to design something on paper that is theoretically excellent but basically impossible to build, either because the shape can't be formed using the tools or materials available or because the design requires very specific tolerances to work.

For example, theoretically a wooden dowel with a 1-inch diameter should fit perfectly into a hole with a 1-inch diameter. In reality, most dowels have a diameter either a little more than or a little less than 1 inch. And when you drill a 1-inch hole, it's almost never exactly 1 inch. Maybe the difference is just $\frac{1}{32}$ of an inch, or maybe the dowel is not exactly round. If your design relies on the dowel rotating in the hole but your dowel is slightly larger than the hole, your design won't work. Likewise, if your design relies on the dowel being held tightly in the hole, but the dowel is even slightly too small, your design won't work. The point is that your design has to take into account the construction method and the reality that everything has a tolerance range.

Another common example is the placement of screws. On a set of shelves that might be secured to brackets by screws from underneath, you need to have enough space below the shelf to allow for the screw length before it is screwed into the shelf and for the screwdriver or drill that will drive the screw—or you need to figure out a way to put those screws in before the shelf brackets are fixed in place. Theoretically the screws might fit, but if you can't actually install them it doesn't matter.

To further complicate things, different materials change dimensions over time and under different conditions. Wood swells and shrinks depending on humidity. Plastic and metal both expand and contract with heat and cold, but at different rates. If plastic tubing is cold when you cut it, and then it gets warmed by the sun, that tubing will become longer by a small percentage.

Over long lengths, this small percentage can add up to inches of difference. You can literally watch warm irrigation drip tape shrink as it fills with cold water. Connecting dissimilar materials is common, but it often leads to joints that loosen over time as the two materials repeatedly expand and contract at different rates with temperature changes.

When I design tools for the farm, I work within a number of constraints, including my own experience and imagination. Most of my designs borrow from other designs I've seen over the years. I not only use materials I've seen used in other tools, I also incorporate ways of connecting pieces, shaping materials, and solving common problems that I've seen work well in other tools.

All of my designs start in my head, but I find it very helpful to sketch them roughly on paper as well. These sketches often point out problems that hadn't occurred to me or identify relationships between parts that could be simplified. I then start turning the sketches into more formal drawings, using a ruler to draw to scale and check the proportions, and getting a better sense of what the tool will look like when it's made from common materials.

I've worked with computer-aided design (CAD) tools that both draw in 2D and model in 3D. These incredibly powerful programs can make very accurate models, and if you enjoy spending your time designing on the computer, I encourage you to do it. But if you only design occasionally and don't want to learn a new computer program, you'll probably find that you can include enough detail in your own drawings using only a ruler or machinist's scale, a simple square, and a compass.

My sketches for simple tools often remain very rough, and I figure out the final dimensions as I make the tool. In a lot of cases this means using whatever scraps of material are handy to put each part together. This approach can work well for one-off tools. But if you're designing something that you want to make more than one of, it is usually easier to draw out the dimensions of everything before you start, and then use standard materials. If someone else is going to build the tool for you—or if you might need to go back and make a copy of a tool a year or two later—this approach is essential.

As you develop your sketches, CAD models (if using), and even your actual construction of tools, it might be helpful to consider that there are two ways to approach your designs: additive and subtractive. These are very much what they sound like. With additive construction, you're adding layers of materials—perhaps gluing or screwing together two separate pieces to make one piece. With subtractive construction, you're carving away the material you don't need—perhaps drilling holes or cutting off unnecessary pieces. You'll likely have many options for when those additions or subtractions take place, and sketching out your design is basically a practice round before you actually start building with real materials.

Basic Math for Tool Design

In both design work and actual construction, basic math—and sometimes even geometry and trigonometry—can be extremely helpful. If you understand basic calculus, you'll be even better off, especially if you're designing tools that have moving parts. You can still get a long way without any formal math, but math is a powerful design tool and having a good grasp of the basics can make your life a lot easier. Throughout this book there are examples of calculations I've made in my design process. Here I explain some of the concepts I use repeatedly.

UNITS

The math you'll use to design tools relates closely to physics. I remember my high school physics teacher drilling into us the importance of including a unit with almost every number we wrote. For example, if we were measuring length we needed to use inches (in), meters (m), or another unit of length. If we were measuring force, the unit would be pounds (lb) or newtons (N), and so on. This is an important practice when making design calculations because you need to know what units you're using, and you need to use the same units throughout each calculation.

Within calculations the only numbers that don't have units are constants, which are numbers like pi (π). Pi is the ratio of a circle's circumference to its diameter and is constant no matter what size the circle is. The decimal approximation of π is 3.142, meaning a circle's circumference is approximately 3.142 times as long as its diameter.

If you pay attention to your units and carry them through as you do calculations, they will also give you helpful clues and reminders, which I'll point out in the following sections. Skip to page 188 to see my notes on making unit conversions.

CALCULATING LENGTH, AREA, AND VOLUME

The concept of **length** is simple, and measuring and adding lengths are some of the most common calculations you'll do when making tools. When you add or subtract lengths, you end up with a new length, and the units stay the same. When using the imperial system, keep in mind that there are 12 inches for every foot, and inches are usually divided into fractions on measuring devices, rather than using tenths, which are the typical divisions that work well with decimal places and digital calculators.

Calculations of **area** come up frequently in irrigation design and occasionally in a few other spots. An area is defined by two lengths multiplied together, and therefore the units of area are a unit of length squared, such as feet^2, also known as square feet. Figuring out the area of a rectangle is a simple matter of multiplying length times width. Remember, though, to use the same units of measure for both length and width. For example, use inches × inches or feet × feet, but never feet × inches.

To calculate the area of any shape that isn't rectangular, you'll still end up with squared units of length, but you can't get there by simply multiplying the length of one side by another. Fortunately, for most farm design calculations you can usually boil down calculations of area to rectangles, triangles, and/or circles (or portions of circles), which I'll talk about in the following sections.

Volume adds a third dimension—depth—to the two-dimensional measure of area. For box shapes with all square corners, the calculation is a simple matter of length × width × height, and the units are length cubed, such as feet^3 or cubic feet.

There are many different units of measure for all sorts of measurements. For example, in farming you'll commonly see square feet (ft^2) as a unit of measure for area, but another common measure of area is acres. (One acre equals 34,560 ft^2.) As a measure of water volume you may run into CCF, which stands for hundred cubic feet, but you'll also see gallons. (There are approximately 7.5 gallons per cubic foot, or 750 gallons per hundred cubic feet.) In addition to the notes at the end of this section on how to use conversion ratios, there's a unit conversion table on page 202 that contains some common units. If you have access to the internet, you can easily convert units by simply asking a search engine.

PRACTICAL TRIANGLES AND BASIC TRIGONOMETRY

Triangles are one of the basic shapes that show up in design calculations. For every triangle, the sum of the three internal angles adds up to 180 degrees. For example, if you know that one angle is 30 degrees and another angle is 45 degrees, the third angle must be 105 degrees because $30 + 45 + 105 = 180$.

In design, you'll repeatedly see right triangles, which contain a right angle (an angle of 90 degrees). In a diagram, a right angle is marked by a small square, as seen in the illustration below. The **area** of a right triangle is half of what the area of a rectangle would be if it had the length and width of the two legs of the right triangle. There are a number of useful calculations you can make for a right triangle, including the **Pythagorean theorem**: $a^2 + b^2 = c^2$, where c is the length of the side opposite the right angle (the hypotenuse) and a and b are the lengths of the other two sides. This allows you to calculate the length of the third side of a right triangle if you only know the lengths of two legs, a common

Basic trigonometry

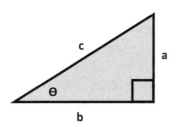

θ is the angle
a, b, and **c** are the lengths of the sides

$\sin\theta = \dfrac{a}{c}$

$\cos\theta = \dfrac{b}{c}$

$\tan\theta = \dfrac{a}{b}$

occurrence when designing parts. For an example of a practical use of the Pythagorean theorem, see page 14.

Right triangles are also fundamental to trigonometry. The basic trigonometric functions (**sine, cosine,** and **tangent**) describe the ratios of the lengths of two sides of any right triangle using one of the other two angles in the triangle. On calculators, sine, cosine, and tangent are usually abbreviated as sin, cos, and tan, and you typically enter the angle in degrees and then hit the button for the function that you want to calculate. Most calculators have a second option of entering the angle in radians instead of degrees, but I've never found this useful. Just make sure that your calculator isn't set to expect radians (usually there'll be a little "rad" in the corner of the screen if it is).

In the old days, before digital calculators, you'd look up the values for sine, cosine, or tangent on tables that gave all three values for every angle between 0 and 90 degrees. Now you can simply plug your angle into a calculator or internet search engine.

So what do those numbers mean? For any right triangle, sine is the ratio between the length of the side opposite the angle you are measuring and the side opposite the right angle (also known as the hypotenuse). Cosine is the ratio of the side next to the angle you are measuring and the hypotenuse, and tangent is the ratio of the side opposite the angle you are measuring and the side next to the angle (this ratio is also known as rise over run). You can do any of these calculations in reverse if you have the ratio and you want to find the angle by using the inverse function, written as $\sin^{-1}$, $\cos^{-1}$, or $\tan^{-1}$.

In designing tool parts that have angles, I frequently know either the angle and length of one side or the lengths of two sides but not the angle. Trigonometry lets me quickly calculate the lengths of the sides or the measurements of the angles that I don't know. I find that knowing the lengths of all sides of a piece is often helpful when calculating how much material I'll need for a project. For example, if I'm cutting angled braces from a 2×4 I might want to know how long the 2×4 actually needs to be on its long side. The 2×4 can be thought of as a long rectangle, and the angled cut essentially makes a right triangle at its end, so I can use trigonometry to calculate how much length the angled cut will add.

When building tools I often find it more accurate, and easier, to measure and mark an angle using a framing square rather than the angle guide on my speed square or a protractor. I can use the tangent function on my calculator to find the desired rise over run for any angle, pick a number of inches on the leg of the square to measure the run, and then multiply that by the tangent of the angle to find the number of inches to measure for the rise.

The Onion Bag Filling Stand (page 114) has numerous angles that I first determined by using rise and run. For most of those angles I also wanted to know the length of the angled side (the hypotenuse) in order to cut matching pieces. It's easy to get the length of the third side using the Pythagorean theorem. As an example, because I knew the length of the bag and the length of the funnel, I was able to determine the length of the ramp piece (the hypotenuse). The height from which I wanted to dump the harvest tote was one of the legs of the triangle. By using the inverse sine, I was able to calculate a good angle for the ramp, which ended up being about 45 degrees.

Trigonometry is also very useful when calculating the magnitude of forces needed. You can see this in action in the section on Mechanical Principles in Plain English (page 191). For example, the magnitude of the torque generated by a force is relative to the sine of the angle between the direction of the force applied and the lever arm. Similarly, in the example of the Mini Barrel Washer (page 128), you can use trigonometry

to find the magnitudes of the forces the barrel exerts on the casters. Due to the angles involved, this turns out to be more than the total weight of the barrel and its contents.

CIRCLES

Circles and parts of circles come up again and again in design. Calculating their dimensions is very basic but worth reviewing here. The distance between the center of a circle and its edge is called a **radius**. The **diameter** of a circle is twice its radius, basically a straight line extending from one edge, through the center, and to the far edge. As mentioned earlier, the **circumference** of a circle (the distance around its outer edge) can be calculated by multiplying the diameter by π (3.142). To calculate the **area** of a circle, you square the radius and multiply that by π. This is commonly written as πr^2.

Circumference of a circle

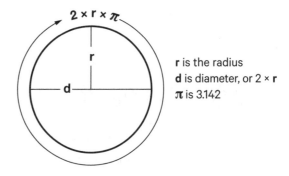

$2 \times r \times \pi$

r is the radius
d is diameter, or $2 \times$ **r**
π is 3.142

Area of a circle

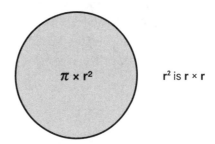

$\pi \times r^2$

r^2 is r × r

CALCULATING AND CONVERTING RATES

Rates are ratios that compare one thing to another. In the section on triangles, I mentioned rise over run, which is the rate at which the rise increases relative to the run. Often rates refer to the amount of time it takes something to happen. For example, gallons per minute describes how many gallons are delivered in a minute.

If the units on the top and bottom of the ratio are the same, it's not necessary to include them in the calculation. For example, 4 inches of rise per 12 inches of run is usually referred to as simply 4:12. If the units on the top and bottom of the ratio are different (such as gallons per minute), you have to include the units for the ratio to be meaningful.

Often you need to convert a measurement in a ratio from one unit to another. For example, if you have a sprinkler that puts out 15 gallons per minute and you're trying to feed it with a hose that allows 480 gallons per hour to flow, you need to change the units of time in one of the two ratios so that you can compare them and see if the two parts are compatible.

A little trick I use to remember how to make conversions to the unit is that conversion ratios have the units you want to convert *to* on one side of the ratio and the units you want to convert *from* on the other side. The unit you want to convert *from* goes on the side opposite of where it currently is.

For example, there are 60 minutes in 1 hour. The conversion ratio can be written $\frac{60 \text{ minutes}}{1 \text{ hour}}$ or $\frac{1 \text{ hour}}{60 \text{ minutes}}$. To convert *from* hours in gallons per hour ($\frac{\text{gallons}}{1 \text{ hour}}$) *to* gallons per minute ($\frac{\text{gallons}}{1 \text{ minute}}$), you use the conversion ratio with hours on the top, opposite the side where it is currently: $\frac{1 \text{ hour}}{60 \text{ minutes}}$. That will give you $\frac{480 \text{ gallons}}{1 \text{ hour}} \times \frac{1 \text{ hour}}{60 \text{ minutes}} = 8$ gallons per minute.

If you wanted to go from minutes to hours, converting the sprinkler's flow rate instead of the hose's flow rate, you would invert the conversion ratio (so it would be $\frac{60 \text{ minutes}}{1 \text{ hour}}$). This approach works for converting any type of unit.

Cost-Benefit Analysis for Tools

IT'S NOT ALWAYS NECESSARY to financially analyze a tool purchase or the cost to build a new tool, but if you're using the tool to make money as part of a farming business, it's important to have a sense of the cost-benefit ratio before you invest in the tool. There are several factors to consider on both the cost side and the benefit side, and they don't all necessarily easily translate into dollar amounts.

Costs

Purchase price. For purchased tools, the cost side of the analysis may seem relatively straightforward: It's the price of the tool, including any maintenance required or "consumables," such as the water used by a barrel washer or drip tape that needs to be replaced every few seasons in an irrigation system.

Build time. If you're building your own tool, figuring out the equivalent cost to a purchased tool is a little more complicated. You definitely want to account for the cost of purchased materials, but you probably should also account for your time. Some of that time will be spent working on the design, and some will be spent building the tool.

Shop time and tools. You might want to account for a bit of "shop time" or include on the cost side of your equation the price of any specific tools you had to purchase to complete a project. If the reality is that the shop would have been there anyway and you would have bought those same tools for other projects, those costs aren't specific to the tool you're building, so it's best to leave them out of your analysis. Otherwise, they'll make the tool look more expensive to build than it really was. Another way to think about this is to consider only costs that you would avoid by not building the tool, and to exclude any costs that you would incur even if you didn't build the tool.

Hidden expenses. There are also costs to building a tool that don't have dollar equivalents. I'm mostly thinking about big-picture environmental costs, but sometimes even costs that seem straightforward, such as some labor costs, are not so easily quantifiable. Examples include the time spent learning how to use a new tool or the need to rework farm systems to use the new tool.

Benefits

Time and yield benefits. You can place a specific dollar value on your time savings by multiplying your hourly wage by the time you save by using the tool. Similarly, the new tool may increase your yield, meaning that the quantity of whatever it is you're producing increases as a result of incorporating the new tool into your system. This, too, is relatively easy to put a dollar value on.

Ergonomic and morale improvements. Benefits that I find hard to quantify, but that I often use to justify building or buying tools, are small ergonomic improvements that don't necessarily speed things up but will result in less damage to my body over days, months, and years of using the tool. Sometimes when I calculate how much I'd spend on a new tool just to "enjoy" a task more,

I realize I'd rather have that money and keep doing the task the old way.

Often, however, I'm okay spending a bit more money in order to work with a tool that makes a task more pleasurable. The point is, I usually do the cost-benefit analysis and make an informed decision. I don't just ignore the costs, but I also don't ignore the less tangible benefits.

Comparing Costs and Benefits on a Time Basis

The life span of a tool is also a factor in my cost-benefit analysis and in how I build tools. Most of the tools in this book are designed to last for many years. When I calculate the operating costs and benefits of a tool, I usually consider its annual costs and benefits, but the purchase price remains the same for the life of the tool. For a rough comparison between two tools, I usually divide the purchase price by the number of years I think the tool will last.

My analysis also includes thinking about what will happen to the tool when it wears out or if I change our systems and stop using the tool in its current form. Will there be a financial or environmental cost for disposal at the end of the tool's life? Or will I likely be able to sell it when it's no longer useful to me?

Ultimately the cost-benefit analysis weighs the costs of a tool against its benefits and helps me decide which one carries more weight.

Tool Analysis Example

To analyze the costs and benefits, I estimate how long it will take for the tool to pay for itself. Any benefits after the tool has paid for itself are returns over that initial investment. Sometimes it might only take a few months for a tool to pay itself off; other times it's a few years. And sometimes the tool will never pay itself off, at least not in easily measurable dollars.

As an example of a simple cost analysis, marking out a bed with the Rolling Bed Marker (see page 54) takes us about 30 seconds per bed on our farm. Our old method of stretching a reel tape and marking lines with a rake took about four times as long and wasn't as consistent. Our labor rate is about $15 per hour, or $0.25 per minute. By using the rolling bed marker, we save about 1½ minutes, or $0.375, per bed.

In a year, we mark beds approximately 120 times, resulting in an annual labor savings of $45. Having our rows and plants more evenly spaced as a result of clearly marking the beds increases our speed planting, weeding, and harvesting, which also results in some savings, but it's harder to put a number on that. My best guess is that those savings are similar to the time saved in the actual marking of the beds.

The parts for the rolling bed marker cost $25, and it takes about four hours to build. With a labor rate of $15 per hour, the total cost to build the bed marker is $85. This means that in about one to two years the initial investment is paid off (depending on whether I choose to factor in the less tangible estimated benefits from planting, weeding, and harvesting rate increases). As long as the bed marker is used as much as I estimate over those one to two years, and it doesn't wear out in that time, it theoretically pays for itself in cost savings during that period.

This is a very rough estimate, of course, and there are other factors that I also consider.

Cost-Benefit Analysis: Rolling Bed Marker

COSTS

Our labor rate is $15 per hour, or $0.25 per minute.

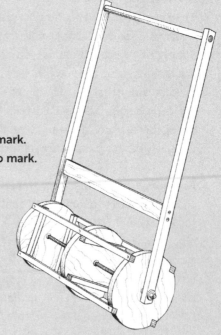

PARTS	TIME (hours × rate)		CONSTRUCTION COST
$25	+ 4 × $15	=	$85

BENEFITS

With the bed marker, each bed takes approximately 30 seconds to mark.
Without the bed marker, each bed takes approximately 2 minutes to mark.
The Rolling Bed Marker saves approximately 1½ minutes per bed.
Labor savings per bed = 1½ minutes × $0.25 per minute = $0.375
We mark approximately 120 beds per year.
Labor savings per year = 120 beds × $0.375 = $45

CONCLUSION

Years to pay off = $85 costs ÷ $45 savings per year = 1.9 years
The Rolling Bed Marker will pay for itself in less than 2 years.

For example, I usually build these tools in the off-season, or at least during times when I wouldn't actually be working out in the field. So, while I do spend time building a Rolling Bed Marker it's something I enjoy doing, and I don't necessarily have to charge myself $60 in labor costs—unless I hire someone else specifically to do field work because I'm in the shop building the tool. Another factor is that I prefer using the bed marker to the former system of pulling lines with a rake—so there's a little benefit there every time I mark out a bed.

On the cost end of the analysis, I consider that the tool takes up valuable space in our tool shed, and sometimes it doesn't mark the beds perfectly. I also constantly come up with ideas for improvements. Because it's going to take up to two years to pay for itself, should I wait two years before investing in something new? Not necessarily.

Once I've invested in a tool, the money is spent. Whether I use that tool until it dies or until I decide I like a different system better, I've already spent the money and I'm not going to get it back (unless I can sell the tool). If a new tool or system will save me even more money than the current tool does, it's probably worth investing in it, as long as it will pay for itself over its expected life.

If I were to give up on the original bed marker after just a year (after initially thinking I'd use it for two years), it's a case of poor foresight, but that may not have been avoidable. In either case, I always base my predictions for how much I'll use the tool on my past experiences and best guesses for the future. I also like to build tools so that they are easy to take apart and rework if I decide to change my systems in the future. It's never really clear to me if the extra labor of taking tools apart is a good financial choice, as it's somewhat labor-intensive, but it always feels good to reuse parts until they are truly worn out.

Mechanical Principles in Plain English

BEFORE I EVER CONSIDERED FARMING as a career I went to engineering school because I wanted to know how things were designed. What I came away with was a better vocabulary to describe many of the natural phenomena I saw when working with materials such as wood, metal, plastic, water, oil, and air—most of which I had been interacting with since I was a young child. Today I rarely do any kind of engineering calculations to check the strength of a part or the stiffness of a structural member, but I have incorporated the vocabulary into the way I think about my designs.

Knowing those concepts, having practiced those calculations, and subsequently having

related that theoretical material directly to my experience in the field—seeing where parts and tools fail and how—has definitely improved my designs. If you want to understand in more depth some of the key concepts I use regularly in this book, read on.

Force
Most of the concepts I'll describe here are ones that you'll find in a good high school physics textbook. **Force** is one of the most basic, and chances are you have a sense of what *force* means even if you don't know this basic formula:

force = mass × acceleration

For most people this formula may be more confusing than clarifying. For example, you know when you push on an immovable wall that it's not accelerating, but you're still exerting a force on it. The reason it's not accelerating is that its connection to the ground, or whatever it's attached to solidly, is pushing back on it with a force in the opposite direction of the force you are exerting on it. When designing tools, it's important to understand that force has a direction, and that more than one force can act simultaneously on a single object. As a result, often forces cancel each other out without creating movement, but they are still there creating stresses or strains on the object.

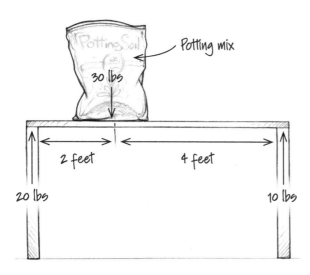

This two-dimensional table with two vertical legs will hold up a bag of potting soil, but because the bag isn't centered on the table, one leg carries more of the weight. (Weight is force generated by gravity.) Each leg transfers a force to the tabletop, and both of those forces together push up with the same force that the bag of potting mix pushes down with. Because the upward and downward forces cancel each other out, nothing moves.

STRENGTH, STRESS, AND STRAIN

On the farm, tools need to withstand heavy use (and sometimes abuse), so it's important to have an idea of how strong your tools will be and how much stress they can take. First, let's look at the difference between strength, stress, and strain. The **strength** of a part is a function of both the material it is made from and the shape of the part. **Stress** is the name for the force on a part relative to the cross-sectional area of the part. And **strain** is the reaction of the material to the stress on it.

If the table legs in the example at left weren't strong enough, they would fail to generate enough force to counter the weight of the potting mix. The strain on the table would be most obvious in the tabletop, which inevitably would bow a little bit under the weight of the bag. As I explain in Bending Forces (page 195), this bowing would happen because the table's top surface would be in compression and therefore would shorten slightly, while the bottom surface would be in tension and would lengthen slightly. The legs themselves would react to the compressive force on them by shortening very slightly and bulging a bit, which reduces the stress very slightly.

If the imbalance in forces is great enough, the potting soil would accelerate down to the ground. Once the potting mix hit the ground, the ground would generate enough force to again counter the weight of the soil. A thicker table leg will have the same force exerted on it but will have less stress on it because that force is distributed over a larger cross section. If you have a tapering table leg, there is more stress on the thinner end and less stress on the thicker end.

When considering forces and stresses on parts, the Mini Barrel Washer (page 128) provides a good example of how geometry affects design. Because the caster wheels provide all of the support for each hoop of the barrel, the distance between the wheels has an impact on how much force is generated internally and how much stress each caster experiences. Setting the two wheels close together would support the weight of the barrel and its contents with less force, while setting the wheels farther apart would make the barrel more stable from side to side but would increase the forces on the wheels (potentially by quite a bit if you take things to an extreme).

Because the caster wheels are able to rotate, they aren't pushing straight up but actually are pushing at an angle. That angle is represented by a line drawn from the center of the caster axle through the center of the barrel. This angled force makes the barrel more stable because the wheel on the right is pushing to the left and pushing up, meaning it's keeping the barrel from being moved to the right at the same time that it's holding the barrel up. The caster wheel on the left is pushing up and to the right with a force equal to the right caster's force.

You can get a sense of the magnitude of those forces by drawing a triangle that represents the vertical and horizontal forces separately. The length of the hypotenuse of that triangle is proportional to the total magnitude of the force exerted by and on the caster's wheel. If the casters were positioned further out to the sides, the angle would be more extreme and the length of the horizontal legs of the triangle (which are proportional to the forces) would be much longer.

Torque

Rotational forces are called **torque** and they depend on the magnitude of the force, on the direction of the force relative to the axis of rotation, and on the distance from the axis of rotation. To understand the concept of torque, imagine trying to loosen a stuck bolt with a wrench. The center of the bolt is the axis of rotation, and to turn the bolt you apply a force to the wrench.

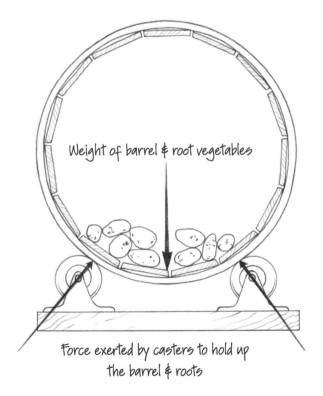

Weight of barrel & root vegetables

Force exerted by casters to hold up the barrel & roots

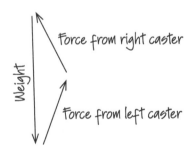

Force from right caster

Weight

Force from left caster

Gravity pushes the barrel and its contents down with a force equivalent to the weight of the barrel and the contents. To keep the barrel from falling, the casters have to push back up with that same force.

When you try to turn a bolt with a wrench, you usually instinctively do two things that maximize the torque you're applying to the bolt: You pull or push on the wrench in a direction that is perpendicular to the wrench, and you put your hand at the end of the wrench farthest from the bolt. This applies the greatest torque to the bolt.

Some people think of this as leverage, but it's more than just the length of the lever. Indeed, if you moved your hand to the middle of the wrench you would apply less force because the lever arm would be shorter. But even if you kept your hand at the end of the wrench for full leverage, if you applied pressure at an angle less than 90 degrees, you would reduce the amount of torque applied to the bolt—even if you pushed with the same force. Both of these actions make it harder to turn the bolt because they reduce your torque.

The formula for torque (represented by τ, the Greek letter *tau*) is:

$$\tau = rF\sin\theta$$

In this equation, r is the distance from the axis of rotation (essentially, the length of your lever), F is the force exerted, and θ is the angle of the force relative to your lever. What this equation shows is that for a given force, you can double your torque by doubling the length of your lever. Likewise, it shows that it's most effective to push perpendicular to the lever and that as the angle changes, the torque is reduced by the sine of the angle.

I think about this torque equation all the time when designing angle braces and gussets to help reinforce joints. A common angle for bracing is 45 degrees, and the sine of 45 degrees is 0.7. This means that to resist rotation on a joint, a 45-degree brace needs to exert 1.4 times the amount of force as if it were perpendicular. The nice thing about using a 45-degree brace for a 90-degree joint, such as between a tabletop and a table leg, is that it works equally in both directions. The brace supports the tabletop, preventing it from rotating around the connection down toward the leg (in other words, from sagging). The brace equally supports the leg, preventing it from rotating toward the tabletop and causing the table to collapse if the tabletop is pushed to the side.

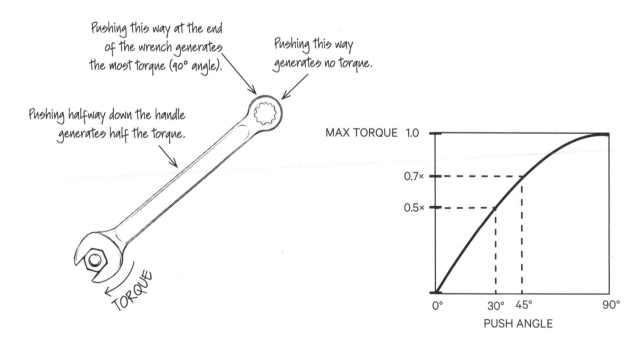

BENDING FORCES

To understand the concept of **bending forces**, imagine a cantilevered tabletop. Assume that the tabletop has a little thickness—maybe it's made from 2×4s on edge, giving it a thickness of about 3½ inches—and it's attached firmly to a very solid wall, but without any braces.

You can probably guess from experience that when you put weight on the outside edge of this tabletop, the top of the edge attached to the wall will start to pull away from the wall. The bottom corner where the tabletop meets the wall will be pushed firmly into the wall. At the same time the tabletop will bend downward, arcing slightly along its length. The important point here is that the top surface of the table is in tension and is being stretched slightly while the bottom surface is in compression and is being smushed a bit. (*Smushed* is not a technical term, but you get the idea; *compressed* would be the technical term.)

Understanding all of these forces and how they are supported is one of the keys to getting the most out of your materials and helping them hold up for a long time. Whenever I make a joint in wood with a screw, I know the screw will start out tight, but if I can use good design to minimize the force on the joint, that screw will stay tight for longer and be less likely to fail over time because the forces on that screw will always be smaller.

Stress Risers

By understanding how a force acts on tools and materials, you can better foresee where exactly a tool will experience higher forces, which cause stress. Remember that stress takes into account both the size and direction of forces, but also the cross-sectional area of the part. You can reduce stress in a part by making the part thicker. However, an abrupt transition between a thick section and a thin section creates a **stress riser** (or concentration of stress) at the transition point.

Stress risers can be advantageous or problematic. An everyday example of an advantageous stress riser is perforated paper. In this case, the "transition" between the relatively thick sheet of paper and the many little cuts in the perforation creates a stress riser along the perforated line. When you pull on the paper—exerting a force that causes stress—the paper fails along the perforation.

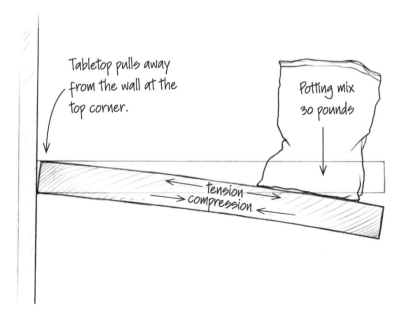

Tabletop pulls away from the wall at the top corner.

Potting mix 30 pounds

tension
compression

The distance of the force on the tabletop from the supporting wall makes a difference: Less distance means less torque, and more distance means more torque and more bending. The thickness of the tabletop will also make a difference in how much it bends because the force creating the tension on the tabletop is farther from the axis of rotation (than it would be with a thinner tabletop), which is the bottom corner of the tabletop.

Sharp corners are examples of less beneficial stress risers. If the sharp corner is on the inside of a part or tool (such as where a table leg attaches to a tabletop), a sideways force on the tabletop will stress the joint between the leg and the tabletop, causing the break to happen at the inside corner. Outside corners are stress risers, too, which explains why they often chip off or wear down more than rounded edges do.

The plywood gussets on the Simple Seedling Bench (page 26) are an example of how to reduce stress risers. Their triangular shape is superior to a square gusset, but the best shape for the gusset to reduce stress risers would be an arcing curve, creating a very gradual transition from the thin leg to the wider gusset. A curve is a less-practical shape to cut, though, and the difference it would make in this application isn't worth the extra effort.

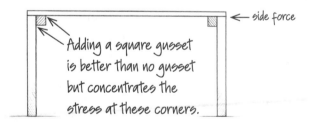

side force

Adding a curved gusset spreads the stress evenly through the arc without concentrating at one point.

side force

Adding a square gusset is better than no gusset but concentrates the stress at these corners.

To greatly oversimplify, you can eliminate, or at least reduce, stress risers by avoiding abrupt transitions in the shape of your parts and the way they connect. For example, you can add an angled brace in the corner of a table leg to help reduce the stress riser in that place.

Fluid Dynamics

In this book I mostly talk about solid materials, but on the farm we also use a lot of water, and it's helpful to understand some basics about fluids. The two fluid-related measurements I most commonly consider when designing systems on the farm are **flow rate** and **pressure**.

Flow rate is usually measured in gallons per minute (gpm) or gallons per hour (gph). Some irrigation parts are noted in measurements of volume other than gallons, such as cubic feet or liters, so you may need to do some conversions in order to complete your calculations. Water commonly flows through pipes and tubing on farms and what gets it to flow, or move, from one place to another is pressure. Most often you see pressure measured in pounds per square inch (psi) or in "bar" in metric units. In gravity-fed water systems and irrigation pumps, the pressure is often measured in units of "feet of head", 1 foot of head is equivalent to 0.43 psi.

The relationship between flow rate and pressure depends on a number of factors, but you can mostly treat them separately for the kinds of tools in this book that use water. The important thing to keep in mind is that you'll likely start with a water source that already has a set operating pressure (most municipal and household systems are set somewhere between 50 and 80 psi). Your water system will also have a maximum flow rate. As long as you don't exceed the maximum flow rate, your pressure will remain fairly constant, probably within about 5 psi. As soon as you exceed the maximum flow rate for your system, you'll notice the pressure drop rapidly.

Force exerted on water to create pressure is generated either by a pump or by gravity. In a gravity-fed system the amount of pressure is determined by the elevation difference between the top surface of the water and the spot where

GRAVITY AND PSI

If you've ever swum to the bottom of a pool, you've felt the pressure gradually increase as you went deeper. At the surface the pressure is essentially 0 psi, and for every foot you go down, the pressure increases by 0.43 psi (1 foot of head = 0.43 psi). If the pool is 10 feet deep, the pressure at the bottom is 4.3 psi. On the farm, if you put a water tank on a hill 50 feet above your field (50 feet of head) and run a pipe down to your field, the pressure will be a little less than 22 psi.

50 feet of head equals about 22 psi.

the water pressure is being measured; this is the direct meaning of *feet of head*.

Most irrigation pumps use impellers. Pumps are designed for an optimal flow rate and pressure. Some pumps are designed for high volume but low pressure, while others are designed for low volume and high pressure.

When you design piping or distribution tubing layout for systems on your farm, there are some important factors to keep in mind. When water flows, its pressure drops slightly, and the faster water moves, the more the pressure drops

(this is called Bernoulli's principle). The motion of water running through a pipe also creates a small amount of friction, and this friction causes additional pressure loss as the water flows through the pipe. The faster the water is moving, the more friction there is, and thus the more pressure loss. In addition, the longer the pipe is, the more the pressure loss from friction adds up.

Anywhere there is a transition in the pipe—such as at an elbow, tee, reducer, or valve—turbulence causes a bit of extra friction. The more abrupt the transition, the more friction loss (pressure loss due to friction) there is. If you're trying to limit pressure losses, make your runs of pipe as straight as possible, avoid elbows if possible, and use gradual bends in the pipe, or even two 45-degree angles instead of a 90-degree angle where possible.

In particular, pay attention to your choice of valves, especially hose valves. Most hose valves are made with a much smaller hole than the internal diameter of the hose or tubing (or pipe) that you're feeding. Look for "full flow" valves that have internal dimensions that more closely match the internal dimensions of whatever hose or tubing the valve is feeding.

Most pressure losses can be kept very small and relatively insignificant by sizing your pipes and transitions appropriately. Even small losses can add up, though, so keep them in mind when you're designing your water distribution systems and when you're troubleshooting problems in the field.

Materials Properties in Plain English

EVERY MATERIAL—including wood, steel, and aluminum—has unique fundamental properties such as strength, weight, and thermal conductivity. Cost and availability are also important properties, although unlike the more fundamental properties, they change over time and by location. What follows is an outline of some basic engineering vocabulary related to materials. Knowing how engineers classify and discuss materials may help you recognize, classify, and evaluate materials for your own designs and discuss options with suppliers.

Ultimately you will learn the most about materials by working with, experiencing, and paying attention to the differences in materials that the objects around you are made from. As you do so, you will generate a catalog in your head of possible materials and their basic properties.

My office bookshelf is still full of engineering textbooks from my college days. In my copy of *Engineering Materials 1*, by Michael F. Ashby and David R. H. Jones, there's a handy table that lists all of the properties and classifies them,

starting with economic properties such as price and availability. Each material also has several mechanical properties, such as density, tensile strength, hardness, thermal fatigue resistance, and creep strength.

The mechanical properties offer information about how a material will react to mechanical forces. For example, how much does the material weigh? Does it absorb vibration? How much force does it take to deform—and to break—the material? Does it hold up to repeated cycles of heating and cooling? Does it deform with small loads that are applied for long periods of time?

In addition, every material has several other properties, including thermal properties; optical properties; magnetic properties; electrical properties; oxidation and corrosion; friction, abrasion, and wear; ease of manufacture, joining, and finishing; and appearance, texture, and feel.

Even if you've never studied engineering properties, if you've been paying attention to the materials around you, you probably have a sense that if you have a steel part and a wood part that are exactly the same size, the steel part will be much, much heavier. The steel part will also be much, much stronger and harder, and it will conduct heat better than the wood part. You might even know that the steel part will conduct electricity better than the wood. You might guess that the steel part will be much more expensive as well.

I use this example to show you that you probably already know a lot about some materials just from experience. By putting numbers on these properties, engineers can make calculations to optimize the material for its intended purpose and minimize factors like cost.

On the farm it's often unnecessary to take the time, and related expense, of calculating what is optimal. In practice, using experience and experimentation to create something that works, even if it's not "optimal," is almost always the way tools and parts get made. The part itself might not be as inexpensive as it could have been if the material choice were optimized with careful calculations. The whole process of making the part for a tool that isn't being mass produced, however, will probably cost less in time savings from avoiding complicated calculations and materials searches. Still, it's helpful to have all of the relevant material properties in mind when selecting materials for your projects.

Some properties will be more important than others for a specific project, but even if you're building something as simple as a table, nearly all of the properties will have an impact on the qualities of the final product.

For example, how heavy is the table? Does setting things on the table produce a loud noise, or does the table bounce sound? Will the table bend before it breaks, or will it break suddenly if it's overloaded? Will the table heat up in the sun and become too hot to touch? Are the table surfaces smooth or rough? How much will the materials for the table cost? What will happen to the table once it is no longer useful? All of these

questions are worth considering when making your initial design decisions.

Basic Types of Materials

There are four basic types of materials: composites; metals and alloys; polymers; and ceramics and glasses. Materials within each of these classes tend to share some basic characteristics, even as each class contains a huge range of materials and material properties. The pages that follow describe a few examples of materials that you'll see in some of the designs in this book.

WOOD AS AN EXAMPLE COMPOSITE MATERIAL

Wood can be incredibly strong, light, and comfortable to touch, as well as aesthetically pleasing, all of which are important characteristics. When compared to some other materials, such as steel or plastic, wood's strength is more directional and variable. The lumber industry tries to reduce the variability of the material, making it more predictable by separating it into different types of wood and different grades, but still there is much more variation from one seemingly similar piece of wood to another than there is with most other construction materials.

Wood is a composite, meaning it's made up of two or more materials in combination. Essentially it consists of long fibers held together with resin, which is a kind of natural glue. The fibers give wood tensile strength—meaning they are strong in tension—but only in the direction that the fibers run.

If you've ever split cord wood, or seen it split, you've seen that splitting the wood along the grain can be done with one or two well-placed hits with a splitting maul. But if you were to use that same maul, or even an axe (which looks similar to a maul but is actually shaped a bit differently), it would take you many, many blows to cut across the log. This is because to split the wood with the grain you only have to pull apart the resin, which is relatively weak in tension, but to cut across the log you have to cut all of the fibers, which are much stronger in tension.

Plywood is an engineered material made by gluing together thin layers of wood, alternating the direction of the fibers in each layer at 90 degrees to the previous layer. This makes it strong in both directions.

Other examples of composite materials are fiberglass, which is made up of glass fibers encased in a polymer resin, and carbon fiber, which is made of carbon fibers encased in high-tech resins. By altering the type of fiber, the type of resin, the orientation of the fibers, and the ratio of the resin to fiber, composites can be made in many ways to address different needs.

STEEL AND ALUMINUM AS EXAMPLES OF METALS

Unlike wood, which has strength that depends on the orientation of the grain, most steel and other metals are equally strong in every direction and are fairly uniform. There is variation depending on the quality of manufacturing process and the purity of the steel, but generally when you buy one piece of steel it will be exactly the same as the piece next to it.

Technically steel is an alloy, meaning it's a mix of different metals, or a metal with another element. Each different metal or other element adds different characteristics to the alloy. There are lots of versions of steel, but all steel is primarily made up of iron and has carbon as part of the alloy.

The most commonly available steel that I use is often referred to as hot rolled carbon steel. Even within this designation there are many variations, but they are largely interchangeable, and you may never know there is a difference unless you ask your supplier. Some of the

differences relate to strength and workability, but often those differences have more to do with how the steel was formed into the particular shape you're buying than how you will work the steel as you make it into something.

Stainless steel is very different from carbon steel. It still has carbon but also contains other alloying metals that make it harder and help it resist oxidation. There are many types of stainless steel.

A commonly used non-steel metal is aluminum. Most aluminum alloys are malleable, meaning they will bend into new shapes and hold those shapes without springing back or fracturing. In general, steel is harder than aluminum, but many types of steel can still be formed into new shapes, just not as easily as most types of aluminum can. You probably recognize this property of aluminum if you've ever noticed how easily aluminum cooking pans bend relative to steel pans.

One of the features of metals is that they tend to conduct heat and electricity very well. Steel is fairly conductive, and aluminum is very conductive. Another feature is that most metals oxidize (react with oxygen). In steel, oxidation creates rust, and for most steels that rust layer has a tendency to flake off, exposing more steel to corrosion. When aluminum oxidizes, it tends to create a hard, protective surface that does not corrode further. Many abrasives are made from aluminum oxide, which is extremely hard. This difference in oxidation properties is why it's more common to see unpainted aluminum and stainless steel parts (chromium in the stainless alloy creates a protective oxide), but most carbon steel is painted to protect it from rusting.

The way that metals conduct and oxidize both have an impact on how easy or hard they are to join by welding. Carbon steel tends to be the easiest metal to weld without significantly changing the strength and corrosion characteristics. Stainless steel and many alloys of aluminum can also be welded, but they require different types of welding processes.

PVC AS AN EXAMPLE OF A POLYMER

In everyday life we tend to lump under the heading "plastics" a lot of materials that basically are all polymers. Again, there is a huge variety of materials and uses within this class. The term *plastic* actually describes a material's ability to be pushed into new shapes without breaking. As an adjective, the word *plastic* applies as equally to metals and other classes of materials as it does to polymers. Obviously glass is not very plastic at all, but metals such as aluminum can be quite plastic. Polymers can be very plastic, in which case they can be shaped easily into all sorts of forms, but they can also be quite brittle.

Polyvinyl chloride (PVC) is a good example of a plastic that sees many uses around the farm. What makes it and other plastics so common is probably that it is very inexpensive. It also has a good strength-to-weight ratio and it is resistant to water and a variety of chemicals. Like many plastics, PVC is not very stable around heat and UV radiation. Different types of PVC contain different additives, which result in different characteristics. Depending on the additives, the PVC can be more or less resistant to both heat and UV, within a range, and it can also be anywhere from very pliable to very brittle.

The most obvious place you'll find PVC on the farm is in the form of pipes. In addition to using PVC pipe as an alternative to metal pipe for carrying water, many farmers also form bows with PVC pipe to hold up plastic film and create all sorts of other structural forms with it because it is so cheap and easy to work with. A lot of

people don't realize that rain gear, boots, and work glove palms are commonly made from fabric coated with PVC. It's in a lot of our electrical cords and plastic tool parts as well.

Both rightly and wrongly, plastic is much maligned by folks worried about the material's ecological impact. PVC in particular is very toxic to manufacture and dispose of, and for that reason I try to avoid using it. On the other hand, part of the reason that it is so inexpensive is that it requires relatively little energy and material to manufacture and transport. When attempting to find alternative materials for an application such as water pipe, calculating how much more expensive a material (such as copper pipe) is does not necessarily show that the alternative is clearly better.

CONCRETE AS AN EXAMPLE OF A CERAMIC AND GLASS

In this class of materials, the features that stand out to me are good compressive strength, relatively low tensile strength, and high density. This is also a category that tends to have relatively low thermal conductivity and good thermal stability. This means that materials in this class don't expand and contract much with temperature changes and that they resist high temperatures. There are of course exceptions, but concrete has all of the characteristics I mentioned above.

On a farm, a concrete floor might be considered an extension of a few of the tools in this book. Concrete floors are excellent for shop spaces as well, and if they are well made and flat, they can provide a good surface on which to lay larger projects that need to be built flat. With good resistance to heat and fire, concrete floors can be a good place to weld and to cut up metal.

ACKNOWLEDGMENTS

AS WITH ANY BOOK, I'm sure there are so many people who have contributed to making this happen in some way that I'm not even aware of all of them. Mostly I want to acknowledge the farmers, farm equipment suppliers, and farm support organizations whose backing and encouragement to build tools for them led me to have enough well-tested designs to write this book. At every step, over more than 20 years of working on farms all across the United States, I've had opportunities to both try out new tools and learn from making new ones.

Going back even further, my parents, and especially the experimental elementary school I went to, made it easy for me to get my hands on tools for building from a young age. One of my earliest memories is using an electric jigsaw to cut out wooden pieces for a toy of my own design—and writing out the plans for it in advance (with the help of my teacher, Mrs. Hudson)!

In an odd way I might also thank the hacker who took down my old, cobbled-together websites, where I originally shared a bunch of the information in this book freely online. The frustration with losing my improperly backed-up posts got me thinking it would be simpler to write a book than to put the website back together properly. (It probably wasn't, and I'll continue to post on the web, eventually.)

METRIC CONVERSIONS

	TO CONVERT	TO	MULTIPLY
LENGTH	inches	millimeters	inches by 25.4
	inches	centimeters	inches by 2.54
	inches	meters	inches by 0.0254
	feet	meters	feet by 0.3048
	feet	kilometers	feet by 0.0003048

	TO CONVERT	TO	MULTIPLY
WEIGHT	ounces	grams	ounces by 28.35
	pounds	grams	pounds by 453.5
	pounds	kilograms	pounds by 0.45

	TO CONVERT	TO	
TEMPERATURE	Fahrenheit	Celsius	subtract 32 from Fahrenheit temperature, multiply by 5, then divide by 9

To convert any of these in the opposite direction, simply reverse the process. For example, to convert millimeters to inches, divide by 25.4. To convert Celsius to Fahrenheit, multiply by 9, divide by 5, then add 32.

INDEX

cutting
  of all-thread, 59
  of foam board, 36
  of pipe, 60
  of plywood, 113
  of slots, 154
  tools for, 12–13, 17
cutting boards, 125

D

data, hacking photos and video
  for, 178–180
data collection sheets, 175–177
dates, 166, 167
designing tools
  basic math for, 184–188
  considerations for, 182–183
  constraints for, 183–184
diagonal bracing, 43
diameter, 187
dimensions, accuracy and, 67
direct seeding, 170
Dramm valves, 125
drill bits, 10, 12, 15, 16
drilling
  holes in 4×4s, 144
  of metal, 16
  overview of, 10–12
  tools for, 15
drip headers, 91, 94–95, 99
Drip Irrigation System, 84–99
drip lines, 91
drip tape, 94, 96, 98–99
Drip Winder, 76–82
drivers, 10–12
Drying Rack, Glove, 108–109
dunnage racks, 120, 125

E

Easy-to-Move Sprinkler System,
  100–105
Ein-Dor sprinklers, 31
elevation, 87
EMT conduit, 34, 51
Engineering Materials 1 (Ashby
  and Jones), 198
ergonomics, 158–159, 189
Excel, 164, 176

F

failure, 183
fatigue and failure, 183
fiberglass, 200
field, tools for
  Hand Cart, 64–73
  Hoop Bender, 48–52
  Rolling Bed Marker, 54–63,
    190
  Snowmulcher, 53
Filling Stand, Onion Bag, 114–
  120, 187
filters, 91, 97
fire safety, 8
fittings, drip irrigation and, 92,
  95, 96
flexibility, spreadsheets and,
  164–165
flow rate, 86–87, 98, 136,
  196–198
fluid dynamics, 196–198
foam board, 36
food safety, 126
force, 187, 191–192
formulas, copying of, 166
Forstner bits, 12
framing squares, 14
friction, 86–87, 197–198
full flow valves, 198

G

gallons per minute (GPM), 90,
  98, 102–103, 136, 196
galvanized pipe, 20
geometry of triangles, 116
Germination Chamber, 34–39
germination rates, 170–171
glass, 201
Glove Drying Rack, 108–109
gloves, safety and, 8
Google Forms, 177
grain, 12
gravity, PSI and, 197
Green Heron Tools, 158–159
greenhouse film, 21
greenhouse tools
  Germination Chamber, 34–39
  Homemade Hoes, 33
  Potting Bench with Mixing
    Tub, 40–45
  Simple Seedling Bench, 26–32
greenhouses
  planting plans and, 170
Griffin, Barry, 53
grinders, 17
gross sales projections, 167
Growing for Market magazine,
  105

H

hacking
  for data, 178–180
Hand Cart, 64–73
hand drills, 12
hand files, 17
hand saws, 13
Hand Truck Pallet, 110–113
harvest plans, 162, 165–168, 169
heaters, 34, 35
hidden expenses, 189

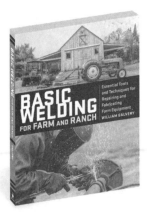